PERFECTIONISM

OXFORD ETHICS SERIES

Series Editor: Derek Parfit, All Souls College, Oxford

THE LIMITS OF MORALITY
Shelly Kagan

PERFECTIONISM
Thomas Hurka

INEQUALITY
Larry S. Temkin

MORALITY, MORTALITY, VOLUME I
Death and Whom to Save from It
F. M. Kamm

PERFECTIONISM

Thomas Hurka

New York Oxford
OXFORD UNIVERSITY PRESS
1993

Oxford University Press

Oxford New York Toronto
Delhi Bombay Calcutta Madras Karachi
Kuala Lumpur Singapore Hong Kong Tokyo
Nairobi Dar es Salaam Cape Town
Melbourne Auckland Madrid

and associated companies in
Berlin Ibadan

Published by Oxford University Press, Inc.
200 Madison Avenue, New York, New York 10016

Library of Congress Cataloging-in-Publication Data
Hurka, Thomas, 1952–
Perfectionism / Thomas Hurka.
p. cm. — (Oxford ethics series)
Includes bibliographical references and index.
ISBN 0–19–508014–9
1. Perfection—Moral and ethical aspects. I. Title. II. Series.
BJ1533.P36H87 1993 171′.3—dc20 92–36601

2 4 6 8 9 7 5 3 1

Printed in the United States of America
on acid-free paper

To the memory of my parents,
in their own ways both perfectionists

Preface

This book's history is both academic and personal.

I first studied philosophy at the University of Toronto in the 1970s, when that department's teaching was still primarily historical. My first exposure to philosophical ethics therefore came through the works of Plato, Aristotle, Spinoza, Hegel, and other classical perfectionists. And the ideas I found in those works fit well with the values I had learned as a child. My parents, themselves products of the liberal, cultured Czechoslovakia of the interwar years, had taught me the intrinsic worth of knowledge, achievement, and the arts, as against mere amusement or material acquisition. In the perfectionist tradition I found a theoretical grounding for my own deepest convictions.

In my last term as an undergraduate and then as a graduate student at Oxford I studied contemporary moral theory, whose techniques I found profoundly illuminating. I decided to apply these techniques to the perfectionism I had found in classical writers but that seemed so thoroughly ignored in contemporary ethics. Two early attempts at this project were theses for the B. Phil. and D. Phil. at Oxford; this book is, I hope, a less inadequate treatment.

For their guidance of my early studies I thank Wayne Sumner, R. M. Hare, Charles Taylor, John Baker, Michael Lockwood, Derek Parfit, and Christopher Taylor. As I began to turn my ideas into a book, intermediate drafts were read by Gavin Lawrence, Donald Regan, Wayne Sumner, Jeff McMahan, the late Flint Schier, and two anonymous referees. Their comments and suggestions all led to fundamental improvements.

In 1989–90 I held an Annual Fellowship at the Calgary Institute for the Humanities to make final revisions to my manuscript. I thank the Institute for its generosity and for providing an ideal environment for my work.

The resulting manuscript received further detailed comments from Donald Regan and from the Oxford series editor, Derek Parfit. Both sets of comments forced me to address some important larger issues I had ignored and saved me from numerous smaller errors. They were models of sympathetic yet forceful critique.

My greatest debts are to Dennis McKerlie and to my wife, Terry Teskey. Dennis, an ideal colleague, has read any number of drafts and discussed countless issues both large and small. His patient encouragement and good judgement have influenced every part of the book. Terry has provided a vigorous philosophical challenge to some of my ideas, and her practised editorial eye has improved every page. Fittingly, a book whose content derives from one family received its final form within another.

Finally, I am grateful for permission to reproduce material from the following: "The Well-Rounded Life," *The Journal of Philosophy* 84 (1987), in Chapter 7; "Why Value Autonomy?" *Social Theory and Practice* 13 (1987), in Chapter 9; "Consequentialism and Content," *American Philosophical Quarterly* 29 (1992), in Chapters 5, 6, and 8; and "Perfectionism and Equality," in Rodger Beehler, David Copp, and Béla Szabados (eds.), *On the Track of Reason: Essays in Honor of Kai Nielsen* (Boulder: Westview Press, 1992), in Chapters 12 and 13.

Contents

PERFECTIONISM

1

Introduction

1.1

Some moral theories have been carefully studied in recent moral philosophy, but one, as important as any, has been largely neglected.

This moral theory starts from an account of the good human life, or the intrinsically desirable life. And it characterizes this life in a distinctive way. Certain properties, it says, constitute human nature or are definitive of humanity—they make humans humans. The good life, it then says, develops these properties to a high degree or realizes what is central to human nature. Different versions of the theory may disagree about what the relevant properties are and so disagree about the content of the good life. But they share the foundational idea that what is good, ultimately, is the development of human nature.

This theory appears in the work of many great moralists. Aristotle and Aquinas think it is human nature to be rational, and that a good human exercises rationality to a high degree. Marx views humans as both productive, because we transform nature through our labour, and social, because we do so co-operatively. The best life, he concludes, develops both capacities maximally, as will happen under communism. For Idealists such as Hegel and Bradley, humans are but one manifestation of Absolute Spirit, and their best activities most fully realize identity with Spirit, as social life does in one realm, and art, religion, and philosophy do in another. Even Nietzsche reasons this way, saying that humans essentially exercise a will to power and are most admirable when their wills are most powerful.

These are just some adherents of the theory; others are Plato, Spinoza, Leibniz, Kant, Green, and Bosanquet. Despite differing in their more specific moral claims, they all offer variants on a single theory, one centred on an ideal of the good life defined in terms of human nature.

1.2

I call this moral theory *perfectionism* and its distinguishing ideal that of *human perfection*. Other terms are available: "naturalism," "humanism," and "eudaimonism" for the theory, "flourishing" and "self-realization" for the ideal. But they have other established uses in ethics and could prove confusing here.

"Perfectionism" has its own disadvantages. If human development admits of degrees, so must human perfection, which initially sounds odd. And some readers may be used to broader definitions of the term. Last century, Sir William Hamilton defined "perfection" as "the full and harmonious development of all our faculties, corporeal and mental, intellectual and moral."[1] More recently, John Rawls has said that perfectionism directs us to "maximize the achievement of human excellence in art, science, and culture."[2] Neither of these definitions mentions human nature, yet each has been influential.

The theory I have identified is perfectionist in both the broader senses used by Hamilton and Rawls. In urging us to develop our natures, it tells us to develop some capacities and also defines an ideal of excellence. My reason for defining "perfection" more narrowly is historical: I think this best fits the usage of writers such as Aristotle, Aquinas, Spinoza, and Leibniz. For them "perfection" means not just excellence, but excellence defined by human nature. The definition also has antecedents in philosophical English. In *Prolegomena to Ethics,* T. H. Green calls his Idealist morality "The Theory of the Good as Human Perfection,"[3] and similar language appears in Sidgwick and Bradley. Sidgwick defines "perfection" as "Excellence of Human Nature,"[4] while Bradley uses "development of human nature," "general perfection," and "perfection of human nature" interchangeably to refer to one moral ideal.[5] My usage mirrors that of Green, Sidgwick, and Bradley: I use "perfectionism" (or "narrow perfectionism") to refer to a moral theory based on human nature, and "broad perfectionism" for the more inclusive view that values some development of capacities or some achievement of excellence.

1.3

This book is a study of narrow perfectionism and, a fortiori, of one form of broad perfectionism. It approaches its subject in the spirit of *moral theory*. According to Rawls, moral theory is "the study of substantive moral conceptions, that is, the study of how the basic notions of the right, the good, and moral worth may be arranged to form different structures."[6] It is descriptive, seeking a systematic account of the moral conceptions people actually hold, apart from questions about their truth or validity. I want to apply moral theory to perfectionism and reach the fullest, most accurate understanding of it that I can. More specifically, I want to use moral theory to arrive at the best or most defensible version of perfectionism. Like other moralities, perfectionism can be developed in different ways, and the results can be more or less plausible. A descriptive account of the best perfectionism is the book's central aim.

Why study perfectionism or apply moral theory to it? One reason is historical: Any view so prominent in our tradition deserves investigation. But there are also moral reasons. I believe that, understood properly and in its most defensible version, perfectionism is an important moral option today. It has often been given inadequate formulations and associated with dubious doctrines. But it can be separated from these, and when it is, has at least three claims on present moral thought.

First, perfectionism has an appealing central idea. That the human good rests somehow in human nature is, although elusive, also deeply attractive. This quality is attested to by its history. If moralists as diverse as Aquinas, Marx, and Nietzsche use the same idea to ground their views, it must have intrinsic appeal.

Second, perfectionism, when combined with a well-grounded theory of human nature, entails attractive particular judgements. Many of us believe that states such as knowledge, friendship, and the completion of challenging tasks are good intrinsically, that is, apart from any satisfactions they may bring. The best perfectionism entails judgements about these states that either match those we already make or take us beyond them in a way we can recognize as progress.

Finally, perfectionism offers to systematize these particular judgements. We initially judge many states to be intrinsically good; a morality that grounds them all in human nature can add coherence and system to our views.

Not only is perfectionism attractive but its study also points to defects in current moral philosophy. On the view now dominant among philosophers, morality concerns only acts that affect other people. It tells us not to frustrate others' desires or interfere with their freedom but says nothing about what we or they should choose for ourselves. Perfectionism strongly rejects this view. It has an ideal for each human—that she develop her nature—and it may criticize her for failing to achieve it. (It may also criticize her for failing to help others develop their nature, but this is not the only criticism it can make.) In my view, its acceptance of *self-regarding duties* is a great strength in perfectionism. A too narrow conception of the moral has impoverished recent moral philosophy and helped limit its influence. Like the great philosophers of the past, reflective people today think there are important questions about how to live one's own life, and they want help answering them. They are rightly impatient with an orthodoxy that ignores this concern. To study perfectionism is to study part of what is most lacking in current philosophical morality.

This point can be put in another way. If the moralities that are currently most studied have an account of the good, it is *subjective,* holding that whether something is good depends on whether it satisfies someone's desires or answers to positive feelings he has. Such an account cannot support serious self-regarding duties, for it excludes any claims about what humans ought to desire. But perfectionism, either broadly or narrowly understood, has an *objective* theory of the good.[7] It holds that certain states and activities are good, not because of any connection with desire, but in themselves. Because its claims about value are objective, they differ essentially from those most canvassed in recent philosophy.

Given philosophers' long neglect of perfectionism, I will spend some time defending it, which will mean drawing out its most attractive consequences and showing how they match our moral convictions. I will also defend perfectionism against some common objections. Some such defence is inevitable even in a work of moral theory. No philosopher would describe at length a morality he did not find plausible or devote as much time to its worst versions as he did to its best. To be worth understanding, a view must appeal to moral belief, and as a result, a descriptive project leaves some room for moral persuasion. My main aim may be to illuminate perfectionism, but I cannot help trying at the same time to place it in a favourable light.

1.4

The book has three parts. Part I (chapters 2–4) discusses the basic perfectionist idea, as far as possible in abstraction from the detailed morality it supports. It asks how "human nature" should be defined; how we can know which properties are contained in human nature; and what, given our best evidence, those properties are. It concludes by defending an "Aristotelian" theory of human nature, one in which human rationality, both theoretical and practical, plays a central role.

Part II (chapters 5–10) elaborates an "Aristotelian perfectionism," one combining the perfectionist idea with the Aristotelian theory of human nature. The first issue it discusses is structure. It argues that the best perfectionism is consequentialist (in a special sense) and agent-neutral, telling us to care not just about our own perfection but about that of all humans. It then expands on these claims, explaining how we seek perfection in whole lives and whole societies, how we compare different perfections, and what specific states contribute to the individual Aristotelian goods.

Part III (chapters 11–13) applies perfectionism to politics, especially to questions about liberty and equality.

Although the book's parts develop a single idea, they are to some extent separable. A reader can find the general account of perfectionism in part I plausible but reject the specific morality developed in parts II and III, either by rejecting my account of its structure or by rejecting the Aristotelian theory of human nature. Conversely, a reader can find the specific morality of parts II and III attractive but not believe it benefits from foundations in human nature. A reader who finds perfectionism plausible only in the broad sense defined by Hamilton and Rawls can read the book's last two parts as developing such a theory, one to which claims about human nature can be added but one that can also be affirmed without them.

Throughout the book I consider perfectionism only as applied to humans. It is possible to generalize perfectionism and to apply its ideal to all living things. Then their good, too, consists in developing their nature. This extension is natural, and the most plausible perfectionism probably accepts it. The claim that human nature explains the human good is deepened if, in general, the nature of living things explains the good of living things. Nevertheless, a generalized perfectionism is beyond the scope of this book. We will consider this theory only if we find its central claims about humans attractive, and there is enough to discuss in them.

The book also examines only *pure perfectionism,* that is, a morality containing only the ideal of narrow perfection and no others against which its claims are weighed. The ideal can also figure in pluralist moralities, in which it combines with other broadly perfectionist values or with non-perfectionist principles about utility or rights. Some may think it a mark of sense to consider perfectionism only in this pluralist context, and I am not unsympathetic to their view. But I think that to understand perfectionism and the contribution it can make to more inclusive views, we should first see what kind of morality it is when taken alone. This is the best way to discover its strengths and weaknesses, and where it does and does not need supplementation. Pure perfectionism may not in the end prove the single most acceptable morality, but it is the strategic subject for perfectionist moral theory.

I

THE PERFECTIONIST IDEA

2

The Concept of Human Nature

In part I, I want to discuss some general questions about perfectionism without getting caught up in details about its structure or implications. For this reason I have defined the theory in a way that is less than completely determinate. I have said that perfectionism is "based on" an ideal of human nature without saying exactly what role this ideal plays. Does perfectionism tell us to seek just our own development of human nature or the development of all humans? How does the perfection in our lives as a whole relate to our achievements at particular times? Questions like these are the subject of part II. Here, at the start of our investigation, it is useful to have a definition that does not settle them either way.

Even a partly indeterminate definition, however, raises questions of interpretation, of which the most important concern human nature. When perfectionism tells us to develop our natures as humans, what exactly does it mean? In what kinds of property does it take our nature to consist? The idea cannot be to develop all our human properties. These are innumerable, and, in any case, many cannot figure in a plausible ideal of perfection. The concept of nature is clearly meant to pick out a subset of human properties, ones that are somehow specially important to being human. Which properties are these?

To develop the best or most defensible perfectionism, we need, most fundamentally, the best concept of human nature. Here there are two tests to apply. Our initial account of perfectionism has moral appeal, and a specification of its central concept must, first, retain this appeal. It must remain close to whatever motivates the idea that the human good rests somehow in human nature, and also reasonably close to the perfectionist tradition. Our nature as defined must seem in itself morally significant. Second, the specification must have intuitively plausible consequences. A perfectionist concept of nature assigns intrinsic value to certain properties, and these must on their own seem morally worth developing. A concept of nature may fail this test by not including some properties that do seem valuable. This flaw is less serious, showing at most that perfectionism needs to be supplemented by other moral ideas. It is more damaging if a concept of nature includes properties that on their own seem morally trivial—if it gives value to what, intuitively, lacks it. This is a telling objection to the concept. A morality based on the concept will be hard to accept because it flouts our particular judgements about value.

Let us give this last objection a name: the *wrong-properties objection*. Then we have a dual task in this chapter. We want to specify a concept of nature that picks out a subset of human properties by using a criterion that is intrinsically appealing and true to the perfectionist idea. We also want a concept that avoids the wrong-properties objection, by having fall under it only properties that seem in their own right worth developing. We can hope that these two desiderata will coincide. If the perfectionist idea is genuinely appealing, the concept of nature most faithful to it should also have the most plausible consequences. Conversely, if a concept of nature picks out wrong properties, it should somehow deviate from the perfectionist idea.

What if the perfectionist idea proves too indeterminate to pick out a single best concept of nature? Then past a point we can only use the second test, about consequences. We can only look at the properties that different concepts select and see which are most attractive. This approach would not be disastrous for perfectionism, nor would it make its talk of "human nature" idle. Not just anything, after all, can count as a concept of nature. Even a partly indeterminate perfectionist idea defines a fairly small area in moral-theoretical space, one that excludes many options. That consequences are then needed to select the best theory within the area does not make it unimportant to have circumscribed the area to begin with. Having to use just the second test would not be disastrous, but I hope to combine it with the first test and show that the concept of nature with the best consequences is also truest to our original understanding of perfectionism.

2.1 Distinctiveness and Essence

What, then, is human nature? Different views are defended in the perfectionist tradition.

2.1.1

One view equates human nature with the properties *distinctive of* humans, or *possessed only by* humans. These properties are important, the view says, because they separate us from other species; no other animal has them. And our good comes in their full development.

Plato suggests this view in *Republic* I, where a thing's good is said to be whatever "it alone can perform, or perform better than anything else."[1] And there are echoes in other writers. Aristotle says human excellence cannot include nutrition or perception because these functions are shared by plants and animals;[2] Kant defines "perfection" as the development of powers "characteristic of humanity (as distinguished from animality)."[3] Even Marx wants to know how humans "distinguish themselves from animals"[4] and tries in several places to show how human labour differs from any activity of animals.[5] So a first view identifies human nature with the properties found only in humans.

Although it appears simple, this view can be difficult to apply. Consider the properties associated with the human digestive system. At the most general level they are shared by other organisms, which leads Aristotle to deny value to their development. Described more specifically, however, they are unique. No other species processes food in exactly the same way as humans or has a digestive system with precisely the same structure. Is there not a distinctive property here? To say no, the distinctiveness view must have some way of excluding specifically described properties, and what this could be is not clear.

Even when it can be applied, the distinctiveness view faces a decisive objection. Humans have some attractive distinctive properties, but they have many others that are morally trivial. Humans may be uniquely rational, but they are also the only animals who make fires, despoil the environment, and kill things for fun.[6] A distinctiveness perfectionism implies that developing these properties is intrinsically good—an absurd implication.

We will see this pattern again: a concept of nature falling to the wrong-properties objection. But in this case the objection's success has deeper roots. If the distinctiveness view fails the second test about consequences it is because it fails the first test, that is, it fails to respect our original understanding of "nature."

Whatever it is, human nature must be something located in humans and dependent only on facts about humans. For this reason it cannot consist in distinctive properties. To say that a property is distinctive of humans is not just to say something about humans. It is to say that the property is possessed by humans *and not by other species*. It is to say as much about non-humans as about humans, and how can facts about non-humans affect *our* nature and *our* good? The point has been well made by Robert Nozick.[7] At present, humans are the only beings with full rationality. But what if dolphins develop rationality? Will this development make its exercise no longer good in humans? What if we discover beings on another planet that have always been rational? Will our rationality never have been good? The distinctiveness view makes the human good depend implausibly on facts about other species, and it does so because its central concept refers to more than humans. It does not capture our intuitive sense of "nature" and, not surprisingly, has absurd implications.

2.1.2

A more promising view uses the concept of essence. It equates human nature with the properties *essential to* humans, or that constitute *the human essence*. "Essence" is used here in the manner of Saul Kripke.[8] For Kripke, an essential property of a kind is one the kind possesses necessarily, or possesses in every possible world where it exists. If a property is essential to a kind, nothing can be a member of the kind and lack it; in every world where the kind exists, its members all possess it. These claims concern necessity, but necessity *de re,* not *de dicto:* that is, they depend, not on conventions of language, but on a kind's own nature. The properties essential to humans are those any being must have to count as human, and the second view says that these properties define the human good.

This view is much closer to the perfectionist idea than was the earlier one about distinctiveness. That a property is essential to humans is a fact only about humans; it involves no other species. Moreover, it seems a fact of just the right kind. A kind's essential properties fix its boundaries or extension; they determine what is and is not a member. Surely this role makes essential properties suitable for a perfectionist concept of nature. If our original ideal was to become fully human or to develop human nature, surely the properties that make us human must figure in its best formulation.

The view also clarifies the appeal of perfectionism. In formulating its ideal, the theory assumes that being human is not just another of our properties, like being a lawyer or a hockey fan. It is fundamental to us. In developing human nature we do not realize something tangential to our identity, but realize what at bottom we are.

The concept of essence clarifies this assumption, for each of us is essentially a human. In every possible world in which we exist, we belong to the human species; in none are we frogs or fish. The essence view therefore makes double use of the concept of essence: to define human nature, and to tie that nature to individual humans. The properties fundamental to the human species are in the same way fundamental to its individual members.

The essence view is well represented in the perfectionist tradition. Hegel calls "universality" the "essence" of human consciousness, and says that only in the society he recommends are humans "in possession of their own essence or their own inner universality."[9] Marx says that capitalism alienates workers from their "human essence," whereas communism will involve "the real reappropriation of the human essence."[10] Similar language appears in Nietzsche. He speaks of "a world whose essence is will to power" and of the will to power itself as "the innermost essence of being."[11] Other writers use equivalent expressions. Aristotle and Aquinas say that goodness consists in actuality or in realizing form, both of which they equate with realizing essence. Still other perfectionists make identity-claims. Plato says in the *Phaedo* that a human is identical to her soul and distinct from her body, from which she is separated before and after earthly life. This statement seems equivalent to a claim about essence. To say that a human is identical to her soul seems to imply that some properties of her soul, such as intellect, are essential to her, whereas those of her body are not.[12]

In its simple form, however, the essence view is too inclusive and falls to the wrong-properties objection. Whatever their other essential properties, all humans are necessarily self-identical, necessarily red if red, and necessarily occupiers of space. None of these properties seems intrinsically worth developing. However well it does on the first test, the essence view fails the second test by including in our nature some intuitively trivial properties.

It may be replied that these trivial properties do not admit of degrees, so including them in human nature cannot affect the important perfectionist judgements distinguishing different modes of living. There may be something to this reply, but I doubt that there is enough. Can we be certain that no trivial essential properties admit of degrees? If humans necessarily occupy space, may some not do so more by occupying more space? More importantly, a concept of nature that includes morally idle properties is, to put it mildly, inelegant. If narrow perfection is a serious moral ideal, it should be specifiable without such useless clutter.

The objections should not make us abandon the concept of essence; it does too well by the first test. We should instead try to narrow the concept of nature so it includes some essential properties and not others: ones that avoid the wrong-properties objection but not ones that are trivial.

2.1.3

The most obvious move is to combine our first two views and equate human nature with the properties *essential to and distinctive of* humans. These are properties possessed by humans in all the possible worlds where they exist and by them alone in the actual world. They are both necessary to and unique to the species.

This essence-and-distinctiveness view has several attractive features. For one, it seems likely to do well against the wrong-properties objection. Because it imposes two conditions for belonging to human nature, it seems likely to select only a few properties and to exclude those most common in wrong-properties objections. Making fires and despoiling the environment are ruled out because, although distinctive, they are not essential; occupying space is out because it is not distinctive. We do not yet know what properties the view positively selects; work would be needed to identify them. But the view seems likely to exclude most trivial properties.

The view also gives probably the truest reflection of the perfectionist tradition. Many perfectionists, among them Aristotle, Aquinas, and Marx, speak both of essence and of distinctiveness, and for consistency must be ascribed a compound view. So, on a different ground, must many others. Writers who claim to value distinctive properties surely cannot mean *all* distinctive properties; the implications are too absurd. They must mean what are distinctive among humans' important properties, that is, among something like their essential properties.

Unfortunately, however, the view inherits defects from the distinctiveness view. It shares the difficulty about what is and is not distinctive, say, about the human digestive system. It also makes our good depend on facts about other animals. Just as a property can cease to be distinctive because of changes in other species, it can also cease to be essential-and-distinctive. Finally, the view pushes perfectionism in a direction I find unattractive. When it works as it is meant to, the distinctiveness test excludes from our good the development of any bodily properties. It supports ideals that are purely mental or even intellectual, such as those of Plato and Aristotle. But these ideals are surely too narrow. That we are embodied animals is a deep fact about us—some would say as deep a fact as any—and one an acceptable perfectionism should reflect. If its aim for us is to develop our nature, surely the bodily parts of that nature must be included. Our physical properties may not be our morally most significant properties or their development our greatest good (see sections 3.1.2, 7.2.1), but a final objection to the compound view is that it gives our physical properties no value at all.

To avoid these defects, we could try amending the view to include only properties that are *essential and necessarily distinctive*, that is, those properties had by humans alone in this world and by humans alone in any world. This amendment helps with the first and especially the second defect. If only necessarily distinctive properties count, we cannot imagine another species's acquiring a property (formerly) in our nature. But it magnifies the third defect, by excluding from our nature properties that many want included. Many perfectionists think rationality is essential to humans, and in its sophisticated forms it is actually distinctive. But it is not necessarily distinctive, for we can imagine other beings with the same rational powers. In fact, we can wonder whether any human essential property, considered on its own, could not in principle be had by another species. If so, the amended view is so restrictive that it lets no properties whatever into our nature.

It seems that no tinkering with distinctiveness will yield an acceptable concept of nature:[13] we need some other device for defining a subset of essential properties. To find this subset, let us look more closely at the different essential properties each human has.

2.2 Essence and Life

Each individual human has six classes of essential properties, distinguished by the range of objects they are shared by. First are the essential properties shared by all objects, such as self-identity and being red if red and, following that, a narrower class found only in physical objects. These properties include being made of (or being) elementary particles and occupying space. Let us identify these two classes as, first, the properties essential to a human qua object and, second, those essential to her qua physical object. Third come essential properties found only in living things and, fourth, those found only in animals. Both these classes contain structural properties. Inanimate matter is made of the same elementary particles as living flesh, obeying the same physical laws. What distinguishes the latter must thus be the particles' organization. To count as animate, matter must be organized for functions such as nutrition, growth, and movement, and what is essential to a human qua living thing or qua animal is that her body is structured for these organic functions. The fifth class contains the essential properties that distinguish humans from other animals, perhaps including rationality. Finally, there are the essential properties that distinguish one human from others. These last properties are essential to her, not qua member of a species, but qua individual, and at the deepest level they are unique to her. Based ultimately on her material origin, they include the particular sperm and egg from which she developed and any further properties deriving from that.

Of these six classes, only the last could not figure in an ideal of narrow perfection. Because this ideal involves a nature common to all humans, it cannot depend on essential differences among them. Nonetheless, individually essential properties may be thought morally significant and deserve some discussion.

2.2.1

Alongside narrow perfectionism there runs, as a secondary theme in our tradition, the idea that humans have individual natures and are better the more fully they develop them. Rousseau says:[14]

> Aside from the nature common to the species each individual brings with him at birth a distinctive temperament, which determines his spirit and character. There is no question of changing or putting a restraint on this temperament, only of training it and bringing it to perfection.

Like the concept of human nature, that of individual nature needs specification and may find it in individual essence. If *individual perfection* develops individually essential properties, is it a value to supplement or even replace that of developing human nature?

Although perhaps initially appealing, the idea does not withstand scrutiny. Consider, first, the fundamental individually essential property, that of being descended from a particular sperm and egg. This property cannot be developed to varying degrees—one either has it or not—and in any case seems morally trivial. Why should facts about a person's material origin affect what is good in her now? Nor do we do better if we consider what her origin brings, namely a continuing genetic endowment. This endowment does not normally give a human the ability to do things others cannot, for example, the ability to do *F* when no one else can do *F*. Instead, it gives her a possibly (but not necessarily) unique profile of abilities to do *G*-to-as-much-as-degree-*m*, *H*-to-as-much-as-degree-*n*, and so on. In itself, this profile cannot be developed to varying degrees; one either has it or not. But it can be used to shape a life. A human can assign time to pursuits in accordance with their place in her profile so that her distribution of activity fits her (possibly unique) configuration of talents. This idea of individual perfection is at least coherent, but it does not have plausible consequences.

If a person's profile contains abilities for fire-lighting and killing, the idea faces a familiar objection. It values intrinsically activities that do not seem of worth. Even if we consider only attractive abilities, there are still implausible consequences. Any broad perfectionism, including human-nature perfectionism, agrees that we should normally be more active where we have more talent. To test the individual-essence view, therefore, we must imagine a case where this will not have the best consequences on other theories of value. Imagine that a person with more talent for music than for writing finds that, because of factors such as the availability of teachers, she can achieve more in writing than in music. Should she still be guided by her profile and give more time to music than to writing? The individual-essence view says yes, but most of us surely say no. For us a profile of talents matters instrumentally, as showing where a person can achieve most, but it does not have intrinsic significance.

An individualist ideal might be plausible if each human's genes gave him a unique style of acting, which could be manifested in all he did and could be developed to higher degrees. Rousseau's talk of a "distinctive temperament" present from birth seems to suggest this idea. But humans do not have genetically based unique styles of acting. Some may develop unique styles, but this is a cultural matter, and contingent. When we consider individual natures as they actually are, there seems little promise in individual-essence perfectionism.

2.2.2

Let us return to our main argument and the five classes of essential property relevant to human-nature perfectionism. Using only the second test, about consequences, which of these classes do we want in our concept of nature? It is easy to eliminate the first class, properties essential to humans qua objects. Shared by numbers and other abstract entities, they are, intuitively, of no moral significance. The same holds for the second class, properties essential to humans qua physical objects. But the remaining three classes—those essential to humans qua living things, qua animals, and (only) qua humans—do seem worth retaining. Considering only consequences, the most promising perfectionism identifies human nature with the essential properties in these three classes. Its ideal is the development of whatever properties are *essential to humans and conditioned on their being living things*. These are essential properties that humans could not have if they were not living; they presuppose life, or are necessarily distinctive of living things

This essence-and-life view retains several virtues from the essence-and-distinctiveness view. It retains part of the answer to the wrong-properties objection because it excludes the trivial essential properties we share with inanimate matter, such as self-identity and occupying space. We do not yet know what properties the view positively selects, and work will be needed to identify these, but the view does exclude many trivial properties. It is also reasonably close to the perfectionist tradition. Unlike the earlier view, however, it does not require difficult decisions about distinctiveness or make our good depend on other species. If some previously inanimate matter acquires a property in our nature, our nature has not changed. Instead, the matter has come alive. Finally, the view recognizes that we are embodied. When fully elaborated, it may not give the development of our physical nature great moral weight—this is a subject for later (3.2.1, 7.2.1)—but it does make it one intrinsic good.

The view avoids certain objections, but does it have a positive rationale? Does our initial perfectionist idea point to just this concept of nature? If not, this is no disaster. The perfectionist idea has taken us a long way, to the equation of human nature with some essential properties. If consequences are then needed to decide exactly which essential properties, this is legitimate fine-tuning of an already substantive ideal. But I believe we can do better. By looking more closely at traditional formulations of the ideal, we can justify this particular specification.

In characterizing perfection, perfectionists speak often of the "good human life." They describe, not a momentary state or achievement, but a whole mode of living. Aristotle, for example, says that perfection can be achieved only "in a complete life"[15] and in weighing the leading accounts of it, compares, not politics and contemplation as such, but whole political and contemplative lives.[16] This emphasis on the life is multiply important. It reflects assumptions about how perfectionist values are aggregated and combined (5.2.2, 6.1–6.3, 7.4.2, 9.2.3), and it also bears on our present concern. The centrality of the "good life" in perfectionism suggests that, whatever properties define it, they must presuppose that we are living. They must contribute to a way of living by themselves being forms of life. Properties shared by inanimate matter are not only intuitively trivial but also irrelevant to an outlook that asks above all how we should live. This outlook can justify its account of nature as follows: We start with the plausible idea that nature is essence and then narrow essence to living essence. This approach keeps our nature within ourselves—there is no dependence on other species—and also fits our original ideal. If that ideal was of a certain human life, we ensure that its elaboration will have an appropriate content.

The justification is deepened in a generalized perfectionism (1.4). If we apply perfectionist concepts to non-humans, we do so only to living things. It is only to animate kinds that we attribute perfection or a nature worth developing. (We do not speak of the good or flourishing of a rock or chemical.) The essence-and-life view reflects this division between kinds in its demarcation of human essential properties. It counts as relevant to our perfection what could be relevant to some species's perfection, and excludes what could not.

The best perfectionism, then, equates human nature with the properties essential to humans and conditioned on their being living things. We cannot yet endorse this perfectionism. We do not yet know what properties *are* essential to humans as living, as we must if we are to decide finally how perfectionism fares against the wrong-properties objection and, more generally, how attractive it is. But, given its account of nature, this perfectionism offers the best hope for a defensible morality. I will underscore this claim shortly by using the account to answer some common objections against perfectionism. But first a clarification is needed.

2.2.3

The perfectionist ideal is a *moral* ideal in the following sense: It is an ideal people ought to pursue regardless of whether they now want it or would want it in hypothetical circumstances, and apart from any pleasures it may bring. In Kant's terminology, the ideal supports categorical, not hypothetical, imperatives, ones that are not contingent on impulses or desires.

The ideal need not be moral in a narrower sense that is sometimes used. In this sense, moral evaluations concern only the choices people make or the traits and dispositions behind their choices. On all theories of human nature, a person's perfection depends partly on her choices, but on many it also depends on factors outside her choices, such as her natural abilities, supply of material resources, and treatment by others. Given such a theory, the perfectionist ideal is moral in the broader sense of supporting categorical imperatives, but not in the narrower sense concerned only with choice and character.

Once it is understood as moral, the ideal can be expressed in several ways. A common formulation is in terms of the "good human life," one in which living essential properties are developed to a high degree. One can also speak of the "good human" (who develops these properties to a high degree) or simply of what is "good" (that the properties are developed). As well, one can speak of what is "good for" a human, if this is defined in terms of the preceding expressions. If something is "good for" a person whenever it is, for example, (simply) good and a state of the person, then developing human nature is "good for" us all. (If I use "good for," and the related expressions "benefit" and "harm," my use will always be in this derivative sense.) But the ideal is not about what is "good for" humans in a more common sense.

In this more common sense, "good for" is tied to the concepts of well-being or welfare and interests: Something is "good for" a person if it increases his well-being or furthers his interests. Well-being itself is often characterized subjectively, in terms of actual or hypothetical desires. Given this subjective characterization, perfectionism cannot concern well-being. Its ideal cannot define the "good for" a human because the ideal is one he ought to pursue regardless of his desires. In my view, perfectionism should never be expressed in terms of well-being.[17] It gives an account of the good human life, or of what is good in a human, but not of what is "good for" a human in the sense tied to well-being.

2.3 Nature: Objections

The wrong-properties objection is the most important against perfectionism (see further 4.4), but there are others. When the concept of nature is undefined, the theory has difficulty answering these objections, but with nature as essence (from now on the restriction to living essence will be assumed) it can respond.

2.3.1

Philosophers used to dismiss perfectionism by saying the concept of human nature is incoherent or obscure. This attitude is less common since Kripke, but a related objection remains: that the concept of nature, although coherent, is partly evaluative.

To include a property in "human nature," the objection says, is not to make a factual claim about it. It is to say the property is somehow important or desirable, and so to use prior evaluative standards. To be a free-standing morality, perfectionism needs a descriptive concept of nature, but this is not available. Its heart, or the heart of any particular version of it, is not an ideal of developing human nature, but whatever criteria guide this prior selection of important properties.

A full answer to this objection requires a positive account of how natures are known (see 3.4). But even here we can see how, with nature as essence, the objection's main claims misfire.

These claims, first, are dubious if they concern the *meaning* of sentences attributing essential properties. The standard analyses of these sentences—and several have been proposed—use the same semantic tools, principally the concept of truth, as analyses of clearly factual language. They contain no explicitly evaluative elements and give no hint of normative content. If the best-known semantics are correct, the concept of essence is descriptive; if nature is essence, so is the concept of human nature.

Second, not just any evaluative content would undermine perfectionism. Only *morally* evaluative content would have this effect, and it is very unlikely.

It is well known that objects can be evaluated from different points of view or by different standards of goodness. Thus, a person can be evaluated first as a musician and then as a potential soldier. Because the criteria relevant to these evaluations are different, the evaluations are logically independent. A good musician may or may not be a good soldier, but his being the one does not entail his being the other. For the same reason, a principle characterizing one good in terms of another—for example, "Good musicians make the best soldiers"—is not damagingly circular. It borrows standards from another evaluative point of view but, relative to its own purposes, these standards are descriptive. It can serve, well or ill, as a ground-level principle for evaluating soldiers.

The same point applies to the evaluation of properties. A principle that uses values to identify the morally important properties of humans will be undermined only if the values it uses are moral. This is unlikely when the properties are picked out as essential.

If we believe, with Kripke, that its atomic structure is essential to an element such as gold, this is not because of a *moral* preference for atomic properties. If anything, it may be because of an explanatory preference. A common epistemological view holds that essential properties are identified by their central role in good scientific explanations (3.4). We know that its atomic structure is essential to gold, the view holds, because this structure is central to the best explanations of gold's weight, colour, and other properties. Even if this view contained the whole truth about essentialism—even if attributions of essence just were claims about good explanation—this would not undermine perfectionism. To say that humans ought morally to develop the properties central to good scientific explanations is to characterize their good by using standards that are not moral. It is to borrow standards from science, and for moral purposes these standards are descriptive. Their descriptive character can, moreover, always be made explicit. If our standards for good explanations demand truth, simplicity, and predictive power, then, on the view we are considering, the perfectionist ideal is equivalent to the following: "The good human life develops to a high degree the properties central to the truest, simplest, and most predictively powerful explanations of humans' other properties." Whether or not this ideal is attractive, it can ground a free-standing morality.

So the concept of nature is not evaluative if nature is essence, and even if it were, this would damage perfectionism only if the concept were morally evaluative. We do not use moral standards to identify the essences of non-human kinds, and there is no reason to think we must use them with humans.

2.3.2

Although perfectionism in general can answer this objection, the same is not true of all versions of perfectionism. In fact, it is not true of many versions. Many perfectionists allow their views about human nature to be shaped by moral considerations and, as a result, make claims about that nature that are false.

This is especially true of perfectionists whose theories are *moralistic*. A moralistic perfectionism takes one human essential property to be something like practical rationality and characterizes this property in such a way that realizing it to a high degree requires developing the dispositions commonly considered virtuous, such as temperance, justice, and honesty, or abiding by the rules—"Do not kill," "Do not lie"—commonly counted as moral. Moralism makes goodness by perfectionist standards in part the same as goodness by the lights of commonsense morality. It makes the degree to which humans develop their natures depend on the degree to which they fulfil popular notions of morality.

Moralism is present in the perfectionisms of Aristotle, Aquinas, and Green, and it dominates that of Kant. In my view it is a fundamental error. Humans are by no means necessarily virtuous, and if moralism implies that they are, it embodies a clear falsehood. Even if it does not imply this strong claim, it embodies a falsehood. Moralism makes claims about the degree to which humans develop their nature that are incompatible with any plausible account of what that nature is. In any sense of "rational" in which it is plausible that humans are essentially rational, it is not plausible that conventionally moral humans are always more rational than immoral ones. Here rationality connotes deliberation and the effective pursuit of ends, which can be found no less in a successful burglar than in a philanthropist. And if there is a sense in which conventionally moral humans are more rational than immoral ones (which I doubt), the rationality it defines is not essential to humans. It is not true that humans could not exist without this morally loaded rationality.

It is one thing to use moral judgements to fine-tune a concept of nature; doing so is acceptable and even necessary (2.1–2.2). It is quite another to let moral considerations affect one's claims about what falls under a concept of nature once that is defined. Moralistic perfectionists, too eager to square themselves with commonsense morality, do the latter and make claims about human nature that, on any acceptable definition of nature, are false.

Nor is this the only defect in moralism. We are interested in perfectionism as a distinctive morality, with distinctive implications. But moralism threatens this distinctiveness. If a moralistic perfectionism says that to develop our nature we must follow such-and-such rules, its consequences will coincide with those of a morality that contains only these rules. The theory's use of perfectionist concepts will still have some importance. It will imply that in following the moral rules, we not only do what is right but also make our own lives best. But this claim will not affect the theory's specific content, that is, its specific claims about what we ought to do. Given this, moralistic perfectionism can seem, not an alternative to other moralities, but at best a variant formulation of them.

This is Sidgwick's reaction in *The Methods of Ethics*. He initially lists perfectionism as one of the main moralities he will discuss, but it quickly loses this status. Having defined "perfection" as "Excellence of Human Nature," Sidgwick says that the principal element in perfection is "commonly conceived" to be virtue, in the conventionally moral sense of virtue.[18] This definition is not surprising, given that Sidgwick's models for perfectionism are Kant and Green, but it has unfortunate results. Because achieving virtue involves following rules similar to those of an intuitionist morality, Sidgwick calls perfectionism a special case of intuitionism and then drops it from his discussion. This is an understandable response to perfectionism as Sidgwick conceives it. Not only does this perfectionism make unsupportable claims about human nature, but it effectively reduces itself to some other morality.

Because moralism is such an error, the best perfectionism must be free of it. It must never characterize the good by reference to conventional moral rules, but always non-morally; and in defending its claims about essence, it must likewise appeal only to non-moral considerations (see chapter 4). Its acceptability will still depend on whether it fits our moral convictions, including some in commonsense morality. But this fit cannot be built into its content by moralistic claims about essence. It must instead exploit features of the theory's structure (see chapter 5).

2.3.3

A second objection claims that, even if the concept of nature is descriptive, the concept of its full development is not. Having identified the properties in human nature, perfectionism must explain what develops them to a high degree, and it cannot do so without appealing to moral considerations outside its main idea.

Even with nature as essence, this objection's first claim is partly true. For some essential properties what their full development involves may be uncontroversial, and for all there are conceptual limits on what degrees of them can mean. But these limits do not always determine a unique measure. Consider, for example, theoretical rationality. How far people develop this property depends presumably on some features of their beliefs, but which features? The concept of rationality itself excludes the idea that the best beliefs are those acquired on a Tuesday or while walking north. But are the best beliefs then true, justified, explanatorily powerful, or some combination of these? To characterize theoretical perfection fully, we do need some decisions, which cannot be justified on narrow perfectionist grounds, about what degrees of it involve.

But this defect in perfectionism is not serious. By the time it specifies degrees, the theory has done considerable work. Its basic idea has selected the properties whose development matters for excellence, and these properties impose substantial limits on what degrees of them can mean. That some room for play is left where further decisions are needed should not make us doubt the importance of perfectionism. It means only that different versions of perfectionism can agree in their central ideal and agree about human nature, but disagree slightly about how that nature is developed. This point will be important later, for in describing the Aristotelian perfections we will often face choices between different principles of measurement (chapters 7–10). But these principles will never be so diverse, nor the play room they exploit so large, that they threaten the theory's distinctiveness.

2.3.4

Finally, there is a weakened version of the wrong-properties objection. It claims, not that perfectionism does have implausible consequences, but only that it could have such consequences if things were different.

Let us concede, the objection begins, that, given the true theory of human nature, perfectionism has attractive implications. Even so, human nature could have turned out to be different. It could have been our essence to be cruel and aggressive, as a caricature of Nietzsche maintains. On this hypothesis, perfectionism has implausible consequences, and this possibility undermines its claim to acceptance in actual conditions.

We could respond to this objection by simply denying, as some philosophers do, that merely hypothetical consequences can count against a moral theory, but this response would not be persuasive. Even if possible consequences count less than actual ones, it is too much to say they count for nothing, especially when the possibilities are realistic and the claims about them dramatic.

A better answer begins by noting that, if nature is essence, there are no real possibilities to consider. The human essence contains the properties humans possess in every possible world where they exist and, on standard assumptions, possess essentially in every such world. If humans are necessarily rational, then, given the most accepted modal logic, they are necessarily necessarily rational.[19] Therefore, when we speculate about what the human essence might have "turned out to be," we cannot be considering a real possibility. We can only be considering what Kripke calls an "epistemic possibility,"[20] that is, the (barely logical) possibility that, given the knowledge we once had, we could have gone on to acquire evidence that the human essence is other than what it is. But this possibility cannot undermine perfectionism. Although it shows that it was once possible for us to acquire the justified belief that perfectionism has repellent consequences, it does nothing to alter the falsity (in all possible worlds) of that belief. If our essence is attractive in the actual world, it is attractive in all worlds, and perfectionism is attractive in all worlds too.

Of course, we can consider the real possibility of there existing beings who are essentially cruel and aggressive, as long as we do not call them humans, but this changes the objection. It does not touch an ideal defined only for humans and not generalized to other species (1.4). Nor does it really damage a more inclusive perfectionism. Given the intimate connection between a kind's essence and its other properties (3.4), essentially cruel beings would have to differ from us in countless other ways. Because their explanatory properties were different, their explained properties would have to be different as well. This makes the earlier skepticism about hypothetical consequences relevant. How can we have reliable moral views about beings so far removed from our experience? How can specific judgements about them outweigh the intrinsic appeal of a morality based on natures? The present objection seems powerful because it seems to show that perfectionism could have implausible consequences for us, living the lives we are now, or for beings very like us. This, however, the logic of essentialism prevents it from showing.

It is not that there is no difficulty here. A generalized perfectionism does imply that, if essentially cruel beings existed, the development of cruelty in them would be intrinsically good. The question is what this implication shows. Given the beings' nature, their perfection would be highly competitive, in that one's achieving it would prevent others from doing the same (5.4). What was intrinsically good might therefore be instrumentally bad because of its effects on the victims. In fact, the situation for these beings' perfection would be bleak: It would be impossible for any one of them to achieve it without directly blocking some others. The generalized perfectionism abstracts from these facts. It allows that a being's exercising cruelty might be instrumentally bad, and even bad all things considered. It claims only that the cruelty, insofar as it realized the being's nature, would be intrinsically good. This is not a claim that, given the beings' remoteness from our experience, our specific judgements give us good reason to reject. And an attractive general ideal gives us reason to accept it.

3

Accretions and Methods

The perfectionist ideal formulated in the last chapter is relatively simple, containing only the concepts of the human essence (suitably restricted) and its development to a high degree. Compared to others in the tradition, this ideal is severely stripped down. Many classical formulations add to it further concepts and doctrines that substantially alter perfectionism's appearance. In my view these concepts and doctrines, which I call *accretions,* are best done without. When present, they divert attention from what is most important in perfectionism and invite needless objections. In the opening sections of this chapter, I identify some historically important accretions to perfectionism and explain why its best version rejects them.

3.1 Accretions

3.1.1

An obviously expendable accretion is the claim that in developing our nature we become more *real,* or acquire more *existence*. Aquinas says that "goodness and being are really the same," and Spinoza says the same about "reality and perfection."[1] Both writers imply that, as goodness increases in degree, so does being or reality, and this claim is explicitly defended by Idealists such as Bradley.[2] But it is hard to see what, aside from rhetorical flourish, it adds to perfectionism considered as a morality. Does any new moral guidance follow from the idea that in developing our natures we gain reality as well as do what we ought? Does the theory acquire new foundations? If not, this strange doctrine should be discarded.

A second accretion concerns *freedom*. It says that true human freedom consists not in any choice among options but in the choice of those options that most develop our nature. Green defines "freedom in the positive sense" as "the maximum power for all members of human society alike to make the best of themselves," and Bradley deems only that human "free" who "realizes his true self."[3] Philosophers may define terms as they wish, but such "positive" accounts of freedom are highly contentious. If they add nothing specifically moral to perfectionism—and again they do not—they should be set aside.

A more important accretion is the claim that developing human nature fulfils our *function* or *purpose* as humans. It appears in perhaps the best-known presentation of perfectionism, Aristotle's in the *Nicomachean Ethics*:[4]

> To say that happiness is the chief good seems a platitude, and a clearer account of what it is is still desired. This might perhaps be given, if we could first ascertain the function of man. For just as for a flute-player, a sculptor, or any artist, and, in general, for all things that have a function or activity, the good and the "well" is thought to reside in the function, so it would seem to be for man, if he has a function. Have the carpenter, then, and the tanner certain functions or activities, and has man none? Is he born without a function? Or as eye, hand, foot, and in general each of the parts evidently has a function, may one lay it down that man similarly has a function apart from all these? What then can this be?

Aristotle goes on to identify humans' function with the development of their essential-and-distinctive properties (2.1.3), in particular, their rationality.

We should confirm the presence in this passage of teleological concepts. Some commentators argue that "function" for Aristotle means "characteristic activity," and in part it does. But we must remember the context of Aristotle's metaphysics. He believes that, of the four causes operating in the natural world, two, the formal cause and the final cause, regularly coincide.[5] It is regularly the purpose of a natural kind to realize its form or nature. This general doctrine lies behind Aristotle's talk of function in the *Ethics*: It is humans' function or purpose to develop their nature because it is the purpose of all living things to do so.

Aristotle sees teleology as a primitive fact about nature, but some perfectionists ground it in theology. Living things, they claim, have the purpose of developing their nature because they were given this purpose when created by God. Their function is not just internal to them, but derived from the divine will.

Whether it is theologically grounded or not, talk of a "human function" is highly contentious; few accept it today. But it is, again, just an accretion to perfectionism considered as a morality. The claim that, in doing what is good, a human also fulfils the human purpose does not alter the theory's account of what is good or right; it has no concrete moral implications. (It may suggest foundations for perfectionism, but they are unwelcome; see 3.2.) For moral purposes, we can reject the claim, and, to develop a metaphysically plausible theory, we should reject it. We should define the perfectionist ideal in terms of human nature without tying that nature to teleological concepts.

3.1.2

Some further accretions are a family of *natural tendency doctrines*. They claim, optimistically, that humans tend naturally to develop their nature to a high degree, and perhaps to the highest degree possible.

One such doctrine concerns *history*. Associated with Hegel and Marx, it says that, as history proceeds, human nature is developed to ever higher degrees, perhaps culminating in a state where it is developed completely. Early in human history, individuals may face serious obstacles to their perfection, which is therefore very limited. As time passes, however, these obstacles fall away, and human development becomes steadily more complete.

A second doctrine concerns desire. It says that humans have a *natural desire* to develop their nature, perhaps as fully as possible. This second tendency doctrine can be qualified by the first, and is so qualified in Marx. He says that workers now avoid "like the plague" the labour that develops their essence, but under communism will regard it as "life's prime want."[6] Other writers affirm the doctrine without time restriction. Aquinas says:

> Each thing is inclined naturally to an operation that is suitable to it according to its form: thus fire is inclined to give heat. Wherefore, since the rational soul is the proper form of man, there is in every man an inclination to act according to reason.[7]

The desire doctrine must be distinguished from a related but weaker doctrine. To remain alive, humans must realize their essential properties to at least some minimum degree. Given this, the familiar doctrine that humans desire self-preservation can be read as holding that humans desire at least this minimum of self-realization. The desire doctrine goes further, saying that humans desire a high or even the highest development of their nature. This is a very different claim. As Nietzsche emphasized, a human who desires the greatest perfection for herself will sometimes risk her life for higher achievements, while one concerned only with survival will not.[8]

Even granting this distinction, the desire doctrine comes in different versions. Weak versions claim only that the desire for perfection is one human desire among many; strong versions claim that it is each human's most important desire. Among the latter, architectonic doctrines claim that the desire for perfection is a supreme organizing desire, one to whose object all other objects are desired as means. ("Means" may include components of perfection.) Whatever humans want, they want in part because it helps develop their nature.

A further distinction is between intensional and extensional versions. Read intensionally, the doctrine says that humans desire objects under the description "state in which my essence is developed to a high degree," or "state in which my F-ness is developed to a high degree" (where F is essential to humans). Read extensionally, it says only that they desire states under some description co-extensive with these. The extensional claim is weaker and, perhaps because of this, more common. Thus, Aristotle and Aquinas affirm doctrines that are architectonic and extensional. They say that all humans have a supreme end to which all others are desired as means, namely, *eudaimonia, beatitudo,* or the good. But, although this end is in fact identical to perfection, many people do not realize this, and, equating it falsely with riches, honour, or pleasure, pursue these goals instead. Their organizing desire will not actually be satisfied unless they develop their natures. But, not knowing this, they do not seek this development as such.

The final tendency doctrine says that developing human nature is a natural source of *pleasure*. Its strong versions claim that perfection is the only source of pleasure, so humans' enjoyment depends entirely on their level of excellence. Spinoza makes this claim necessarily true by defining pleasure as "the transition of a man from a less to a greater perfection."[9] Weaker claims are also possible. One (not mentioning essence) is Rawls's "Aristotelian Principle," which says that, "other things equal, human beings enjoy the exercise of their realized capacities (their innate or trained abilities), and this enjoyment increases the more the capacity is realized or the greater its complexity."[10] Rawls does not claim that (broadly) perfectionist activity is the only source of pleasure or that its enjoyments are always more intense than those available elsewhere, but he does say that realizing human capacities is, as such, always pleasant.

In any version, the pleasure doctrine may have implications for desire. If humans desire pleasure, as many believe they do, it entails at least a weak extensional desire doctrine: For some of humans' desire for pleasure to be satisfied, they must develop their nature. If humans realize this, they will have some tendency towards perfection.

3.1.3

Because the tendency doctrines are factual doctrines, they can be only accretions to perfectionism. But many perfectionists accept them, with unfortunate effects on how their moralities are understood.

Imagine that you are a perfectionist and accept a strong desire or pleasure doctrine. Then there is a life that develops human nature, that you think is good, and that you also believe involves satisfaction. ("Satisfaction" here refers to pleasure or desire-fulfilment.) As a perfectionist you believe this life is good only because it develops human nature and not because it involves satisfaction, but your acceptance of the tendency doctrine may obscure this. Noting that your ideal involves pleasure or desire-fulfilment, some may think your morality is in a familiar way satisfaction-based, with just an idiosyncratic view about where satisfaction is found.

Some perfectionists warn explicitly against this misreading. Aquinas says that, when we know God's essence and achieve our highest perfection, we will also enjoy "delight" and "repose of the will." But he insists that the knowledge "ranks before" the delight, which is merely "something attendant on it" and devoid of value.[11] And Bradley says:

> Pleasure is an inseparable element in the human end, and in that sense is necessarily included in the end; and higher life implies pleasure for the reason that life without pleasure is inconceivable. What we hold to against every possible modification of Hedonism is that the standard and test is in higher and lower function, not in more or less pleasure.[12]

Such warnings, however, are less frequent than they might be in the tradition, and their absence has led many perfectionists to be sometimes interpreted as utilitarians. Despite the contrary indications in their works, Plato, Aristotle, Marx, and Green have all been read as concerned ultimately with pleasure or desire-fulfilment.[13] To avoid this misreading, we must separate the tendency doctrines from perfectionism and insist that at best they are accretions to it.

3.1.4

The question then remains whether any tendency doctrines are true. If so, perfectionism, although distinct from any satisfaction-based morality, is extensionally equivalent to one, supporting the same judgements about right and wrong.

Surely no doctrine strong enough to support this equivalence is true. History has no single tendency, nor do all humans have one unifying desire. Some desire perfection but many do not, and even those who do desire perfection often have competing desires that are stronger. As for the pleasure doctrine, some weak version of it seems plausible. Perfectionist activity is often enjoyable, and we may therefore affirm a weak version of Rawls's Aristotelian Principle.[14] But we cannot affirm a strong version. Although perfectionist activity is one source of pleasure, it is not the only source or always the greatest source available.

In rejecting the doctrines, we need not go so far as Nietzsche, who associates perfection not with satisfaction but with suffering:

> You want, if possible—and there is no more insane "if possible"—*to abolish suffering*. And we? It really seems that *we* would have it higher and stronger than ever. . . . The discipline of suffering, of *great* suffering—you do not know that only *this* discipline has created all the enhancements of man so far.[15]

Instead, we may hold that perfection involves just moderate losses in pleasure or desire-fulfilment. This is Green's view. He says "there is pleasure in all realisation of capacity," so "the life in which human capacities should be fully realized would necessarily be a pleasant life." But this life has no monopoly on pleasure, and the frustration of "lower tendencies" it involves may result in a smaller aggregate of pleasure than in a worse life.[16] This view seems reasonable. Because humans can desire and enjoy developing their capacities, a perfect life will usually be reasonably satisfying. It just need not always be the most satisfying life possible.

Even this moderate view may seem disturbing. Some philosophers argue that morality must concern satisfaction, yet a perfectionism without tendencies can recommend lives that are not the most satisfying.

Far from damaging perfectionism, this divergence is a source of strength. Imagine a research scientist. If she is dedicated to her work, she will experience frequent frustration as well as sometimes the thrill of successful discovery. Her desire always to be advancing knowledge—a desire not even the best satisfy—may make her life less contented as a whole than if she had simpler wants. Do we then think her wrong to pursue science? Would her life be better with more easily satisfiable desires? Surely not. Her scientific talent is what is best in her and what she should most strive to develop. Or consider the lives in *Brave New World*. They are extremely satisfying, on any plausible account of "satisfying," yet lack perfectionist goods such as knowledge and autonomy. If morality had to concern satisfaction, these lives would have to count as ideal, yet we all find them repellent.

This point can be made differently. If a strong desire or pleasure doctrine were true, pursuing excellence would be easy. Once we knew where our greatest good lay, achieving it would be just a matter of following our strongest want or enjoying our greatest pleasure. This is not our experience. For most of us, achieving the good requires discipline and concentration. It requires formulating a valuable project and sticking to it despite distractions and temptations. Given this, perfectionism without tendencies not only matches our moral convictions but also fits our experience of seeking a valuable life.

My claim is not that satisfaction has no value. Pure perfectionism makes this claim, but there is also the possibility of a pluralist theory that weighs perfectionist ideas against others about, for example, pleasure or desire-fulfilment (1.4). Such a theory can combine these ideas in different ways. It can treat satisfaction as simply another value alongside perfection, or it can say that satisfaction has value only,[17] or has the most value,[18] when it is satisfaction *in* perfection, for example, pleasure in scientific research. But it is no objection to a pluralist theory that it does not treat satisfaction as the only value. It may not even be an objection to pure perfectionism that it does not treat satisfaction as any value. (I am uncertain about this.) It is, however, a decisive objection to a pure satisfaction-based morality that it does not treat perfection as a value—this makes the morality unacceptable.

3.2 Perfectionist Naturalism

To argue that perfectionism is acceptable is to defend or try to justify it. How is this done?

3.2.1

Many philosophers assume that a successful defence of perfectionism will involve *naturalism*, in the meta-ethical sense of naturalism. It will not justify moral claims by connecting them to other moral claims, either more particular or more general. Instead, it will derive them from non-moral facts. It will argue that facts about human nature directly entail conclusions about the human good. Thus, if it is essential to humans to be rational, it follows logically that humans ought to develop rationality. This could be so only if the general perfectionist ideal, which affirms the value of developing human nature whatever it contains, were conceptually or necessarily true. On a common interpretation, perfectionism treats its ideal as having this status.

This interpretation has some support in the tradition, for some perfectionists do appear to be naturalists.[19] But their meta-ethical view is at best an accretion to perfectionism. One can be a perfectionist and a naturalist, trying to derive one's values from facts, or one can be a perfectionist and an anti-naturalist. There are good reasons to prefer the second approach.

The best argument against naturalism turns on the intimate connection between evaluation and action. As Hume noted, evaluations are action-guiding in a way that mere descriptions are not. To assent to a descriptive statement, one need only form a certain belief, say, that humans are rational, but to assent to a moral judgement one must do more. One must act as the judgement directs, or at least form the intention so to act when circumstances are appropriate. If one agrees verbally that humans ought to develop rationality but does not do so oneself, then, if there is no special explanation, one's assent was not sincere.

This difference between evaluation and description refutes the central claims of naturalism. If an evaluative term such as "good" is action-guiding, it cannot be defined in purely descriptive terms without losing an essential part of its meaning. And no argument with purely descriptive premises and an evaluative conclusion can be valid, for its conclusion contains an element not present in its premises.

Although it is often used, this argument makes a psychological assumption. It assumes that humans are free to act in different ways or to pursue different ends, and it would have no force if this assumption were not true. In particular, it would have no force against perfectionist naturalism if a strong intensional version of the natural desire doctrine were true (3.1.2). Then each human's strongest desire would be for perfection *as* perfection, and her assent to ''x will develop your nature'' would already have connections to action. She would already want to pursue x, and this association would carry over to ''x is good,'' if the two claims were synonymous.

In fact, given a strong intensional desire doctrine, a positive argument for perfectionist naturalism could be developed. If each human necessarily desired just her own perfection, there would be no room for morality to prescribe some other end: If it tried to do so, its prescriptions would be idle. The only possible role for morality, given the fixity of human desire, would be to prescribe means to the one desired end, as perfectionism does.[20]

This argument for naturalism has only theoretical interest, given that we have rejected any strong desire doctrine (3.1.4), but it helps in understanding the tradition. Many perfectionists who appear to be naturalists accept desire doctrines and may derive their naturalism from them.

Consider, for instance, Aquinas. He says that being (or perfection) and goodness are ''really the same,'' a claim naturally read as asserting a necessary connection between them. But his argument for the claim turns explicitly on desire:

> Goodness and being are really the same, and differ only in idea; which is clear from the following argument. The essence of goodness consists in this, that it is in some way desirable. Hence the Philosopher says (*Ethic.* i): *Goodness is what all desire.* Now it is clear that a thing is desirable only in so far as it is perfect; for all desire their own perfection. But everything is perfect so far as it is actual. Therefore it is clear that a thing is perfect so far as it exists; for it is existence that makes all things actual.[21]

This argument is hardly pellucid, but a key premise is that ''all desire their own perfection,'' a clear desire claim.

Similar reasoning appears in Bradley. In *Ethical Studies,* he tries to show that ''self-realization'' (his name for perfection) is the one moral end, and to do so naturalistically. Once we ''know what we mean, when we say 'self', and 'real', and 'realize', and 'end','' he writes, we will see that self-realization is our end.[22] But his supporting argument, again, concerns desire. He sketches a theory of desire in which the object of desire is always the agent's greater perfection, or, in Bradley's language, in which ''what is desired must in all cases be self.''[23] Given this theory, the pursuit of any end other than perfection would be ''psychologically inexplicable,''[24] and moralities requiring it can be ruled out of court.

Finally, a similar argument is implicit in Aristotle's talk of the ''human function.'' For Aristotle, a final cause comes last in a natural sequence of events.[25] Given the coincidence of formal and final causes, this means that humans have a natural tendency to develop their nature, reflected in a desire at least extensionally directed at this end (3.1.2). But if our natural tendency is towards perfection, what else could our good consist in? In fact, Aristotle's talk of "function" expresses an approach to the formulation of perfectionism that has been very prominent in the tradition.

3.2.2

In this traditional formulation, the development of human nature is each human's "end," in three senses. It is his metaphysical purpose, perhaps given him by God; it is what he actually strives for or desires; and it constitutes, by logical necessity, his good. These three uses of "end" are connected. That perfection is a human's purpose helps explain why he pursues it: all things tend naturally towards their purpose. (Perhaps the God who assigned the purpose also instilled the tendency.) And its being what a human strives for prevents anything else from being his good. I call this formulation *teleological,* because of its triple use of "end," which encapsulates a *teleological* view of human nature.

This teleological formulation is common in the tradition and has been the subject of many critiques. But its claims about metaphysics, psychology, and meta-ethics are all accretions to perfectionism. They do not affect the theory's substantive claims about what is good and right and should therefore be abandoned. The best perfectionism defines its ideal in terms of essential properties without tying those properties to any metaphysical purposes. It says that humans may desire perfection but denies that there is any psychological compulsion to this. And it denies that its ideal is conceptually true. The ideal must be defended, not by logic, but by substantive moral argument.

In this stripped-down version, perfectionism does not have the grand ambitions ascribed to it by writers such as Alasdair MacIntyre and Bernard Williams.[26] It does not aim to show the incoherence of moral nihilism or to provide some logically inescapable route into morality. It does not claim to find in human nature an "Archimedean point" from which morality as a whole can be rationally justified. The writers who ascribe this ambition do not think it can be realized, because it requires a teleological biology that is no longer scientifically credible. On my interpretation, however, perfectionism does not aim to "justify" morality.

This aim is impossible, first, if by "morality" we mean something like commonsense morality, a familiar code requiring other-regarding virtues such as justice and beneficence. MacIntyre and Williams do understand morality this way. They seek from perfectionism a justification, in terms of an agent's own good, of the conventionally required treatment of others. As I have argued, however, a serious perfectionism must be serious about human nature, and on no admissible concept of nature does a conventionally good human always develop human nature more than a conventionally bad one (2.3.2).

The aim is also impossible if "morality" means just a willingness to act on ideals that may conflict with present desire. On my interpretation, perfectionism is not a magical entrée into morality, but a substantive position within it. It assumes a general willingness to act on moral ideals and proposes a specific ideal to follow. Some philosophers find perfectionism intriguing because they think it has grand ambitions, and teleological formulations may encourage this thought. But the ambitions are chimerical for any morality, and we do perfectionism no service by considering it only in the context of an impossible philosophical project.

3.3 Defending Perfectionism

3.3.1

A substantive defence of perfectionism must follow the same lines as a defence of any morality. It must show that the theory coheres with our intuitive moral judgements at all levels of generality, or, in Rawls's phrase, is in "reflective equilibrium"[27] with all these judgements. Because perfectionism is complex, a full defence of it must be complex. It must show, first, that the general perfectionist ideal, that of developing human nature, is attractive when considered by itself and as a potential foundation for morality. The theory's basic idea must have intrinsic appeal. Second, the defence must show that the ideal has attractive consequences, both at the middle level, where it identifies the properties whose development is intrinsically good, and at lower levels, where it makes particular claims about which activities are best and right. Finally, perfectionism must work as a systematic whole, with its general ideas explaining its particular claims. Then the theory not only matches our intuitions but also gives them a satisfying rationale.

This kind of defence is most persuasive when there is some logical distance between a morality's parts, so that judgements about one are not infected by knowledge of the others. There often is this logical distance within perfectionism. Because it takes argument to identify the properties essential to humans, we can evaluate the theory's general ideal without knowing what specific values it supports and, conversely, have views about the value of developing, say, rationality without knowing whether rationality is part of human nature. Similar logical gaps appear elsewhere in the theory. A general ideal of rationality does not reveal immediately what specific activities realize it best, and there will be further confirmation for the theory if these activities turn out to be attractive. Because perfectionism's parts can be considered in relative isolation from each other, its passing reflective equilibrium tests is especially impressive. If the most plausible concept of nature turns out to contain only independently attractive properties, with independently attractive realizations, the theory is independently confirmed at all its levels.

The defence is made easier by the fact that, at each more general level, perfectionism can make adjustments to avoid unwanted consequences at lower levels. It can refine its concept of nature to exclude morally trivial properties (2.1–2.2) and, in specifying degrees of perfection, can prefer measures with more attractive implications (2.3.3, chapters 6–10). But the room for adjustment is limited. Not just anything can count as a concept of nature, and not just anything can define degrees of rationality. Throughout perfectionism, the more general levels impose constraints on what can be affirmed at levels below. If despite these constraints the theory's particular claims are still attractive, this supports its general ideal.

The second and third parts of the defence turn on perfectionism's consequences. In the remainder of this book, I will argue that, in its best version, perfectionism has many attractive consequences and does not have unacceptable ones, as it would do if it assigned value to what are clearly trivial properties (see further 4.4). Ideally, the theory would yield every attractive consequence or capture every intuitively appealing moral claim. This may not happen; there may be intuitively appealing properties that cannot be connected to an ideal of human nature. (Perhaps the capacity for pleasure is one.) If so, this failing is less damaging to perfectionism than if the theory valued intuitively trivial properties. At most, it shows that the perfectionist ideal must be combined with other moral ideas in a pluralist theory (1.4). And whether it shows this depends partly on the first part of the defence: whether the perfectionist ideal is attractive when considered by itself.

If the ideal was not attractive by itself, its plausibility would depend entirely on its ability to yield and explain attractive consequences. That there were desirable consequences it could not capture would suggest strongly that it needs to be supplemented in a pluralist theory. If the ideal is attractive by itself, however, the situation is different. Then we may hold it against an initially appealing moral claim that it cannot be connected to human nature. We may even reject the claim, deciding that it is mistaken if it cannot be connected to human nature. This issue is one we can discuss here: the intrinsic appeal of the perfectionist ideal.

3.3.2

I believe the ideal does have intrinsic appeal: The goal of developing human nature, or exercising essential human powers, is deeply attractive. This is reflected in its widespread acceptance. The ideal is implicit in non-philosophical talk of living a "fully human" or "truly human" life and is endorsed by diverse philosophers. Writers such as Aquinas, Nietzsche, Green, and Marx have very different particular visions of the good life. Some value contemplation; others value action. Some value a communal life; others value a life of solitude. If, despite these differences, these philosophers all ground their particular values in a single ideal of human nature, that ideal must have intrinsic appeal.

Some of this appeal can be explained, if nature is essence. Because each of us is essentially a human, to develop human nature is not to develop some temporary or tangential property, such as being a lawyer or a hockey fan (2.1.2). It is to develop what makes us what we are.

The ideal is also attractive as a potential foundation for morality. It may not exclude a justification in terms of more basic principles; perhaps no moral idea does. But it does not require such a justification. It is of sufficient depth and generality to be in itself the basis of all moral claims; in human nature we have something that can be ethical bedrock. In fact, the perfectionist ideal makes a peremptory claim: As a potential foundation for morality, it dismisses any moral judgements that cannot be derived from it. However appealing they seem in themselves, it counts against them that they cannot be connected to human nature.

It may be objected that these claims for the ideal are less plausible if we have rejected the accretions discussed earlier in this chapter. If developing human nature was each human's purpose, given her by God, it would make sense that doing so was good. Stripped of these claims, the ideal has less appeal.

There may indeed be some cost to abandoning these accretions. If each human had a metaphysical purpose, the claim that it was good to fulfil that purpose might have intrinsic appeal. If so, a perfectionism that talked of purposes would have two attractive basic ideas, one about human nature and one about purposes, and in rejecting the accretion we lose the support of the second idea. But the loss here is not great: The perfectionist ideal, considered apart from any accretions, is attractive enough to ground a moral theory. And there may be no loss at all. Do teleological perfectionists believe first that it is good to fulfil the human purpose, and only then, because perfection is our purpose, that perfection is good? Or do they believe first that perfection is good, and then, because our purpose must be good, that our purpose is perfection? I suspect their reasoning often follows the second course. If it does, teleological claims add nothing to the appeal of perfectionism.

The same is true of accretions about a divine creator. That an all-powerful being wishes us to pursue some goal is no reason to do so unless we know, independently, that what the being wishes is good. Unless perfection is intrinsically desirable, someone's desire that we pursue it has no moral weight. This point is reflected in the perfectionist tradition. Theological perfectionists such as Aquinas do not derive the perfectionist ideal from claims about God; they apply it to him. If God is supremely good, they hold, it is because he fully realizes his unlimited nature. In addition, there are important non-theological perfectionists. Marx and Nietzsche are resolutely irreligious and have a purely mechanistic view of nature. Nevertheless, both endorse an ideal based on human nature.

It may also be said that perfectionism loses by rejecting the natural tendency doctrines. If a strong desire or pleasure doctrine were true, perfectionism could explain why some moralists equate the good with satisfaction: They mistake something strictly correlated with the good for the good itself. Again, there may be some loss here. But if a moderate view like Green's is true (3.1.4), perfectionism can give some explanation of these moralists' error: Satisfaction, although not strictly correlated with perfection, does often go with it. And even if Green's view is false, it is hardly a decisive objection to a moral theory that it must sometimes say that other theories are simply false.

The intrinsic appeal of the perfectionist ideal is not seriously diminished by the rejection of various accretions. We have good reason to examine the substantive moral theory it generates.

3.4 How Are Essences Known?

To elaborate this perfectionism, we must first determine what properties do belong to human nature or actually are essential to humans and conditioned on their being living things (2.2.2). This inquiry in turn requires a method for identifying these properties. Deciding which essential properties are conditioned on humans' being living seems straightforward; the difficult question is how we discover which properties are essential. How in general are essences known?

3.4.1

Epistemological questions have received less attention than one might have expected in the recent literature on essentialism, but philosophers who defend particular claims about essences use one of two methods.

The first method, associated with Kripke,[28] is *intuitive*. It says that we discover essential properties by making intuitive judgements in thought experiments involving candidate members of a kind. To learn whether its atomic structure is essential to gold, for example, we imagine a series of possible substances with gold's atomic structure but a different outward appearance. If we judge all these substances to be gold, our judgement shows that its inner constitution is essential to gold and its phenomenal properties contingent. We learn what could and could not be gold by asking how we could and could not imagine gold's existing.

The second method examines *scientific explanations*. Associated with Hilary Putnam,[29] it says that we identify essential properties by their central role in the explanations given by good scientific theories. That gold has a certain atomic structure explains its colour, weight, and other phenomenal properties, but is not in turn explained by them. Gold's atomic structure is thus explanatorily prior to these properties, and this shows, on the second view, that it is essential. For the explanatory view, properties central to scientific explanations are essential, and in gold these cluster around its inner constitution.

These two methods agree in their conclusions about many particular essential properties, for example, those of gold. But they are very different, and the choice between them is important for epistemology. The choice is also difficult. Each method faces difficulties, and in each case they threaten to reduce it to its rival.

The intuitive method faces the standard objection to any appeal to "intuitive" knowledge. Do we really have direct intuitions about *de re* necessity? Are our judgements about possible objects really self-standing? In this particular case the objection is especially pressing, for no one can make these judgements without having some collateral knowledge. No one can know that a kind has a property essentially without knowing first that it (simply) has that property. Thus, no one can know that gold has its atomic structure essentially without knowing first that it (simply) has that structure. Moreover, one may need further knowledge. Of the intuitive method we must ask: If people knew all the properties of gold but did not know their explanatory relations—whether its structure explains its appearance or vice versa—could they make confident judgements about possible instances of gold? Could they decide which were and were not gold? If not, this would undermine the intuitive approach. Our supposedly free-standing verdicts in thought experiments would reflect prior explanatory knowledge, and the intuitive view would collapse into the explanatory.

The difficulty facing the explanatory method appears when we try to analyze "explanation." The best-known account of explanation is Carl Hempel's "deductive-nomological" account,[30] but it will not do for identifying essences. It notoriously fails to capture certain "asymmetries of explanation," of which one is that essential properties explain accidental properties but are not explained by them. If there is a deduction of gold's phenomenal properties from its atomic properties plus some bridge principles—which is all the account requires—there is also a deduction of its atomic properties from its phenomenal properties plus some (reversed) bridge principles. Explanations are supposed, on the second view, to distinguish essential from accidental properties; as analyzed by Hempel, they cannot.

To capture the asymmetries of explanation, we must supplement Hempel's account with further material conditions. Some philosophers have proposed the following: A deduction counts as explanatory only if it derives accidental properties from essential properties and not vice versa.[31] But this proposal undermines the explanatory approach. If we need to know which properties are essential before we can know which deductions are explanatory, we cannot use explanations to identify essential properties. On the contrary, we must have some prior, perhaps intuitive, knowledge of essential properties. This most obvious supplement to the deductive-nomological account undermines the explanatory method, and it is not clear whether any other supplement gives the right results about explanation.

Given these difficulties, I cannot take either method and show that it provides the one canonical means for discovering essences. Nonetheless, it seems that *at least one* of the methods must be canonical. And, since whichever is not canonical seems likely to collapse into the other, it may not matter practically which is which. If we are interested in a practical question such as "What is human nature?" we can proceed as follows. We can use both methods simultaneously, testing our claims about the human essence both intuitively and against explanatory theories. We can count as essential to humans whatever properties are picked out by both methods together. With gold these properties seem to coincide, and, if the previous arguments are correct, they should coincide generally. Epistemologists may differ about which method is doing the real work in our arguments and which is dependent, but if the same conclusions follow from both methods, then adherents of both should find them well grounded.

3.4.2

Against this double approach, an objection may be raised. How can we use the second, explanatory method if we have rejected all strong tendency doctrines (3.1.4)? If there was some property that all humans wanted to develop, or that history was tending to develop, the method would be easy to apply. A teleological human science would give this property a central role, and explanatory considerations would then show that the property was essential. Without tendencies, however, no teleological science is possible, and in the absence of such a science, how can the second method be used?

A teleological science would, if true, bear on questions about our nature, and many perfectionists do use such a science to ground their essentialist claims. Aristotle's belief that formal and final causes coincide implies that we discover a kind's nature by seeing what it naturally grows towards. Hegel and Marx make essential to humans the very properties whose development is the end state in their theories of history, and Nietzsche says:[32]

> Suppose, finally, we succeeded in explaining our entire instinctive life as the development and ramification of *one* basic form of the will—namely of the will to power, as *my* proposition has it; suppose all organic functions could be traced back to this will to power and one could also find in it the solution of the problem of procreation and nourishment—it is *one* problem—then one would have gained the right to determine *all* efficient force univocally as—*will to power*. The world viewed from inside, the world defined and determined according to its "intelligible character"—it would be "will to power" and nothing else.

This epistemic use of teleology is, in fact, central to the teleological formulation of perfectionism (3.2.2). This formulation claims that, if humans naturally desire some property, its development must be part of their good. It can claim on the same ground that the property must be part of human nature. The fact about desire gives the property a central role in explaining human behaviour and thereby shows it is essential.

Although the explanatory method can use a teleological science, it need not. Think again of gold. No scientific theory says that gold tends to develop its atomic structure to higher degrees; such a theory is impossible. What our best science does, rather, is to use gold's atomic structure to explain the many different changes it can undergo. If gold is heated to a certain point, it melts; if it is immersed in aqua regia, it dissolves; and in each case the explanation starts from the same facts about its atomic structure. Something similar is possible for humans. Even if their behaviour does not always aim at one goal, there can be a single theory that explains how they pursue their different goals, and a single property that is central to all the explanations this theory supports. Even if human action is diverse, its explanation can always start from the same property. If so, explanatory considerations will make this property essential even though it is not the goal in any teleological science.

4

The Human Essence

We are now in a position to ask what properties *are* essential to humans as living things. Using the intuitive and explanatory methods, this chapter will defend what I call an *Aristotelian theory of human nature*. According to this theory, humans share with other animals certain bodily essential properties but are also essentially rational, in both the theoretical and practical senses of "rational." Together with the perfectionist ideal, this theory yields an *Aristotelian perfectionism,* one with three values: physical perfection, which develops our physical nature, and theoretical and practical perfection, which develop theoretical and practical rationality.

4.1 The Aristotelian Theory: Physical Essence

4.1.1

A first deliverance of the intuitive method is that humans necessarily have bodies. We can imagine purely spiritual beings and, perhaps, understand their psychology. But if they have no physical form they are not, intuitively, of our species. Beyond this, intuition tells us that humans necessarily have bodies with a fairly determinate structure. No human can remain alive without a functioning respiratory, muscular, digestive, circulatory, and nervous system, and, analogously, no possible being without these systems passes the intuitive test for humanity. Unless its body permits it somehow to breathe, move, process nutrients, and exercise central control, it is not a human.

These claims are confirmed by the explanatory method. One explanation of human behaviour is physical explanation, and it makes central the very systems picked out by intuition. To explain why a runner is panting, we say her circulatory system needs to carry more oxygen to her muscles, which causes her respiratory system to process air at a greater rate than usual. If she pushes off with her legs, we say her nervous system is sending messages to her thigh muscles, causing them to contract. These explanations supervene on and may reduce to the explanations of some more basic science such as chemistry or physics. But this reduction is irrelevant in our version of perfectionism. We have defined human nature to exclude any properties shared by inanimate matter (2.2.2), which means that the explanations we look to must likewise be restricted to living things. Chemical and physical explanations apply to rocks and gases as well as to humans, and we must therefore consider only explanations that presuppose some organic structure. In humans this structure is expressed in certain major physiological systems—in respiratory, muscular, digestive, circulatory, and nervous systems—and the relevant explanations cite their operation.

Both methods have difficulty making these initial claims more precise. If humans essentially have some respiratory system, must they have the specific arrangement of organs they have? If we imagine beings like us in all respects except that they have three lungs, are we imagining humans? What of beings whose respiratory systems operate on different chemical principles? These questions are difficult, and I cannot answer them decisively, but they may not be important in a moral study of perfectionism. In our world, no human can grow a third lung or alter chemical laws; we are stuck with the physiological systems we have. Given this, our physical perfection can depend only on how our actual systems function, that is, on events in our actual bodies. If a general description of our physical essence—one saying only that we need some respiratory system—defines clear degrees of our physical perfection, it may not matter whether more specific descriptions do or do not remain essential.

4.1.2

What, then, defines degrees of physical perfection? This is more a question for physiology than for philosophy, but a rough answer is as follows. Each system in our body has a characteristic activity. The respiratory system extracts oxygen from air, the circulatory system distributes nutrients, and so on. For a human to remain alive, each system must perform its activity to some minimal degree; for her to achieve reasonable physical perfection, it must do so to a reasonable degree. But a system does this when it is free from outside interference and operating healthily. So the basic level of physical perfection is good bodily health, when all our bodily systems function in an efficient, unrestricted way. Then essential physical processes occur to a reasonable degree, and we have reasonable physical perfection.

This first implication is attractive. Even apart from their effects on other values, illness and poor organic functioning are intrinsically regrettable. The loss of a limb or of full activity in an important organ detracts from the completeness of a human life, and robustness adds to it. It makes the life more fully human. Physical health may not be a major perfectionist good, and it may receive less moral weight than the development of rationality (7.1.1). But a perfectionism that gives it some value acknowledges that some of our nature is bodily (2.1.3) and that, like others, this part can be more or less developed.

This connection with health is no accident, but has evolutionary origins. Like other aspects of our nature, our bodily systems were selected as those most likely to make for our survival and reproduction. Their unimpeded operation is healthy because otherwise beings possessed of them would be adaptively disadvantaged. In their present form the systems are essential to humans, who cannot exist without them. But, like humankind itself, they emerged from natural selection and were favoured over alternatives precisely because of their connection with healthy, self-maintaining activity.

Higher physical perfection comes in vigorous bodily activity. Here our major physical systems perform to higher degrees, processing more air, carrying more nutrients, and moving greater weights longer distances. This activity occurs most notably in athletics, and Aristotelian perfectionism finds the highest physical good in great athletic feats. These feats often embody perfections other than physical perfection. They require skill and dexterity and can follow months of careful planning. In both these ways they realize practical rationality, and this can account for much of their value. But there is also a physical dimension. When a human runs 100 meters in 9.86 seconds or long-jumps 29 feet, something physically splendid occurs. His bodily powers are realized to the full in a way that is intrinsically admirable and of intrinsic perfectionist worth.

Most of us are not outstanding athletes and cannot achieve the highest physical perfection. Still,we can preserve our basic health and pursue whatever mild athletics are compatible with our main projects. We have instrumental reasons to do both these things. Physical activity keeps us alert and can be the medium for some exercise of rationality. If Aristotelian perfectionism is correct, however, this activity is also a modest intrinsic good, as the development of our physical nature.

4.2 The Aristotelian Theory: Rationality

4.2.1

The most important Aristotelian claim is that humans are essentially rational. Its elaboration requires an account of degrees of theoretical and practical rationality, which will be given later (chapters 8–10). But its core is this: Humans are rational because they can form and act on beliefs and intentions. More specifically, they are rational because they can form and act on sophisticated beliefs and intentions, ones whose contents stretch across persons and times and that are arranged in complex hierarchies. These last features distinguish human rationality from that of lower animals. Animals have isolated perceptual beliefs, but only humans can achieve explanatory understanding. They can grasp generalizations that apply across objects and times and can use them to explain diverse phenomena. A similar point holds for practical rationality. Animals have just local aims, but humans can envisage patterns of action that stretch through time or include other agents and can perform particular acts as means to them. By constructing hierarchies of ends, they can engage in intelligent tool use and have complex interactions with others. Distinctive properties do not matter as such in our perfectionism (2.1), but the Aristotelian theory makes essential a kind of rationality that at present is found only in humans.

That humans are essentially rational is supported, first, by the intuitive method. We do not think there were humans in the world until primates developed with sufficient intelligence, and the same view colours our judgements about possibilities. If we imagine a species with no capacity for a mental life, or with none more sophisticated than other animals', we do not take ourselves to be imagining humans. Whatever their physical form, they are not of our species. The degree to which humans exercise rationality varies from time to time in their lives, being lower, for example, when they are asleep. But beings who never envisage or plan for a future are not, intuitively, humans.

Again, this claim is confirmed by the explanatory method. Alongside the physical explanation of human behaviour is psychological explanation, which explains at least intentional human action by citing beliefs and aims that make it rational. This explanation presupposes theoretical and practical rationality and ascribes their current exercise in all its particular accounts. It deserves closer examination.

4.2.2

A psychological explanation of person *A*'s act of ϕ-ing has the following form:

> *A* intended to make it the case that *p*.
>
> *A* believed that ϕ-ing was the most effective means to *p*.
>
> *A* was acting as a rational agent.
>
> *A* was physically able to ϕ.
>
> Therefore, *A* ϕ-ed intentionally.

Two features of this schema deserve comment. One is that it begins with an intention that *p,* rather than going back to a desire or wish that *p*. This point highlights the central role of intentions in practical reasoning. We have many desires that we never act on, because their objects are unattainable or because we think satisfying them is on balance unwise. These desires never affect our behaviour. Only when a desire generates an intention do we think seriously about its satisfaction or set ourselves properly to pursue it. The second feature is the claim that *A* was acting as a rational agent. This claim is required for the success of any psychological explanation but, within that explanation, is always contingent. Even if humans are essentially rational, they do not exercise full rationality at every moment. On the contrary, they sometimes succumb to weakness of will, self-deception, and other lapses from full rational control. To be subjects of psychological explanation, they must be generally rational, and generally do what their beliefs and aims make appropriate. But full rationality need not always be present, and, when it is not, full rational explanation is not possible.

In this general form, psychological explanation applies to some other animals who also act on beliefs and aims.[1] This is not so, however, when the ascribed mental states are sophisticated, with extended contents and hierarchical relations. Then the premises of the explanation—beliefs about scientific laws or intentions for the distant future—are beyond other animals, and the conclusion may be as well. It may involve intelligent tool use or willed co-operation with others. The scope of the explanation also alters given sophisticated mental states. Instead of taking *A*'s intention that *p* as given, sophisticated psychology can say that she intends *p* as a means to *q*, which she in turn wills as a means to *r*. It can explain a particular end as a means to others that appear above it in a rational hierarchy. It can also explain many beliefs. If *A* believes that ϕ-ing is the most effective means to *p*, sophisticated psychology may say:

A has evidence that ϕ-ing is the most effective means to *p*.

A is forming beliefs as a rational agent.

Therefore, *A* believes that ϕ-ing is the most effective means to *p*.

The rationality ascribed here is no longer the practical rationality that derives acts from intentions and beliefs, but the theoretical rationality that grounds beliefs in evidence. It is the rationality exercised when sophisticated beings use general principles to move from initial evidential beliefs to other, more speculative beliefs. The explanation does not claim that *A* is perfectly rational or always forms beliefs on the basis of evidence. She may sometimes suffer from slips of reasoning, wishful thinking, or self-deception. For her beliefs to be generally explicable, however, she must generally derive them from evidence, and for this particular explanation to succeed she must be doing so now.

The content of sophisticated psychology, then, gives it a sophisticated form. Instead of treating items of behaviour one by one, it ascribes a system of connected beliefs and aims to explain, not just a person's acts, but many of the mental states behind them. It explains aims in terms of beliefs and other aims, and beliefs in terms of other beliefs. The central role in this psychology is clearly played by the properties of theoretical and practical rationality. These properties are, first, presupposed in every premise of a psychological explanation. No one can have a belief or intention about *p*, especially if *p* is sophisticated, without the mental capacity to grasp its content. In saying that she believes or intends *p*, we assume that she has that degree of rationality. Nor can she have a belief or aim about *p* unless she generally acts on her beliefs and aims as reason requires. Unless she generally does what she believes will promote her goals, no goals can be ascribed to her. Unless she generally derives beliefs from evidence, she cannot have beliefs. Finally, a premise ascribing the present exercise of rationality is a crucial part of every psychological explanation. Unless *A* is now acting or forming beliefs rationally, the premises of the explanation will not entail its conclusion. Unless her rational powers are now being exercised, her having certain mental states will not explain a thing.

That the two forms of rationality are central does not, however, imply any natural tendency doctrine (3.1.2). Sophisticated psychology does not say that humans have an overriding desire to develop rationality or a supreme tendency in that direction. On the contrary, it places no restriction on the content of their goals. What it does, rather, is use the one property of rationality to explain how humans' different goals all issue in action. Its structure is therefore like that of explanations of gold. It starts from one ascription of rationality and shows how, given this rationality, people with different initial aims, experiences, and evidence will end up believing, intending, and acting differently. It uses reason to explain goal-directed behaviour without making reason itself a goal.

Psychology also makes rationality essential without supporting moralism, the view that developing one's own nature requires the other-regarding virtues (2.3.2). Because the rationality it ascribes is formal, defined only by the scope and interrelations of a person's beliefs and aims, it can be realized as much in conventional immorality as in morality. If the immorality is wrong, it cannot be because it reduces the agent's own perfection (see further 5.2.3).

4.2.3

This first explanatory argument gives us good reason to conclude that humans are essentially rational, but it is not the only such argument available.

As described to this point, sophisticated psychology uses pairs of beliefs and aims to explain human acts. How does it decide which pairs to use? How does it know which specific mental states to ascribe? In the case where *A* ϕ's intentionally, we may say that she intended *p* and believed that ϕ-ing was a means to *p*. But it would also explain her act if she intended some different end *q* or *r* and believed that ϕ-ing was a means to that, or intended not-*p* and believed mistakenly that ϕ-ing would prevent *p*. To give a determinate explanation of her ϕ-ing, psychology must be able to select one of these intention-belief pairs above the others. How does it do so?

The only answer I know makes rationality even more central to sophisticated psychology. Several writers, among them Donald Davidson, argue that this psychology is governed by a "principle of charity" requiring us to make an agent's overall behaviour as rational as possible.[2] In explaining a particular act, we must ascribe those beliefs and aims that make the most sense of her conduct as a whole. Different intention-belief pairs may do equally well in rationalizing her present act, but they will differ in their capacity to fit into a larger scheme explaining her total conduct through time. One allows us to find rational origins for most of her other beliefs and acts; another forces us to leave many ungrounded in reasons. The principle of charity exploits this difference and, by requiring us to prefer the first ascription, permits a determinate explanation of what she has done.

Without something like a principle of charity, it is hard to see how psychology could give determinate explanations. But again, some clarifications are needed.

The principle of charity does not imply that everyone's behaviour is highly rational. It tells us to interpret for maximum rationality, but it cannot say what result this effort will have. One person's conduct may be such that its most charitable interpretation makes him very rational, whereas another's leaves him, even on the kindest construal, much further down a scale of coherence. An ideal of charity guides psychological explanation, but it cannot determine its final content. That depends on empirical facts about a person's behaviour.

Second, a plausible principle of charity has two parts, one theoretical and one practical. It tells us to maximize both theoretical and practical rationality, or rationality in belief and rationality in action. These different maximands can sometimes conflict, as in cases that suggest self-deception. To say that someone has deceived himself is to explain an act of belief formation as a rational means to some goal, for example, avoiding distress, but it is also to ascribe an unjustified belief. In deciding whether to interpret someone as self-deceived, we decide whether to maximize his theoretical or practical rationality, where we cannot do both.

A final point concerns the substance of the ideal of charity. When sophisticated psychology tells us to maximize theoretical and practical rationality, what exactly does it intend? What is the precise content of its interpretive goal? According to a simple view, rationality involves just acting on some intentions and believing on some evidence, so a charitable interpretation maximizes just the number of a person's acts and beliefs with some rational origin. But this view is not the only one possible. Davidson, for example, says that interpretations should make a person so far as possible "consistent, a believer of truths, and a lover of the good."[3] This proposal goes beyond the simple view because a belief based in evidence may still be false and an act aimed at an end far from laudable. In fact, Davidson's account of charity is just one of many possible. On the theoretical side, we can imagine a weak principle telling us to maximize just the number of a person's beliefs that are consistent; two stronger principles telling us to maximize either the number that are justified or the number that are true; and a still stronger one telling us to maximize the number that are both justified and true, or that constitute knowledge.

There are further possibilities. Charity can tell us to prefer ascribing beliefs that are more sophisticated. It can, and in my view should, say that, other things equal, we should prefer ascribing beliefs with more extended contents and more elaborate inter-relations. We make agents more rational, and explain them better, if we assign states with greater reach and explanatory coherence. Unfortunately, views like this last one are hard to defend. However plausible it is that psychology requires *some* concept of charity, it is difficult to argue that this concept must take one rather than another specific form. Can we show that one account of charity is truer than all others to our everyday explanatory practice? Or that it gives what by independent criteria are clearly better explanations? I do not see a decisive argument here, and it may be that there is none. It may be that, despite the general importance to psychology of using some principle of charity, there is no one content that principle must have.

If this is right, there may be some indeterminacy in the concept of charity, but the indeterminacy does not undermine the general conclusion we can draw from Davidson. Even without a specific principle of charity, we can know what range an acceptable principle must fall within, and we can also know its importance. If sophisticated psychology tells us to maximize the rationality (in some sense) of all agents, then rationality (in that sense) is even more central than before. Rationality is presupposed in the premises of every psychological explanation, and its current exercise must also be explicitly ascribed. Now we see, beyond this, that it is a regulative ideal governing psychology and determining which of the many rationalizations consistent with its general structure are indeed explanations. If psychology uses charity constraints, rationality is doubly central to it, and we have doubly good reason to include rationality in the human essence.

Alongside physical perfection, then, Aristotelian perfectionism recognizes two further goods, which we can call *theoretical* and *practical perfection*. Although their full description will come later, they already look promising, and a perfectionism containing them seems likely to have attractive consequences. This is impressive, because in deriving the goods from the human essence we did not use moralism or any moral arguments. The claim that humans are essentially rational first emerged from thought experiments and then was confirmed by psychological explanations.

When this is added to the intrinsic appeal of the perfectionist idea, the prospects for Aristotelian perfectionism look good.

4.3 The Aristotelian Theory: Objections

Like any theory about so contentious a subject as human nature, the Aristotelian theory faces objections. Let us consider some of these and then return to the most important objection to perfectionism, the wrong-properties objection.

4.3.1

Some may argue that, by emphasizing rationality, the Aristotelian theory ignores the emotional side of human nature. This objection may be directed to the theory's factual claims or to the attractiveness of a morality based upon them. Either way, it rests on an untenable contrast between rationality and emotion. Rationality is a property of beliefs and aims, which are present in all emotions. Our fears and loves have as components beliefs about dangers or the merits of our friends, as well as desires to avoid or spend time with them. Like other beliefs and aims, they can be more or less rational; when they are more rational, the emotions they help constitute are by Aristotelian standards good. (See further 10.3, 10.4.1.)

4.3.2

A more far-reaching objection challenges the whole idea that rationality can be essential, saying the explanations it figures in are not serious explanations. Although in common use today, rational psychology will eventually be replaced by physical and especially neurophysiological explanation, which will account for all human behaviour in purely physical terms. People may still talk casually of beliefs and aims, but neither these mental states nor explanations citing them will be viewed as scientifically serious.

This objection raises large issues that cannot be settled finally here, but let me sketch two responses to it. Joining for a moment in confident assertions about the future of science, I do not see how neurophysiology can explain human behaviour without finding some physical analogue of rationality. If its laws link stimulations, say, of the retina with behaviour, they cannot do so directly, for humans do not always react to visual cues in the same way. Sometimes when humans see food they eat it, but sometimes, when they are not hungry or believe the food is poisoned, they do not. So neurophysiological laws must mention other physical states corresponding to hunger or the lack of it, belief that the food is poisoned, and so on. They must also find a structural property of the nervous system that explains why some combinations of states produce eating behaviour and others do not. This structural property, however complex, will correspond to rationality, and it will play the same central role in neurophysiological explanations that rationality plays in sophisticated psychology. Reference to it will be a constant in explanations of what we call intentional acts, and the assumption that it is having its normal effect will be crucial for the explanations' success. If neurophysiology eventually explains all human behaviour, it may lead us to replace one conception of rationality with another. But it cannot stop our believing that humans somehow possess rationality, and possess it essentially.[4]

This response concedes that neurophysiology may one day replace psychology as we know it, but this point too can be resisted. Some philosophers hold that psychological and physical explanations are independent, each operating by its own criteria and having its own validity. Intentional human acts admit of two explanations, neither of which reduces to nor is replaceable by the other. Davidson defends this view on the basis of claims about charity. He argues that psychology's use of charity constraints gives it a holistic structure not found in the physical sciences and prevents an explanatory reduction to them.[5] If this view is defensible, psychology may not be threatened by any developments in neurophysiology. No matter how much we learn about the brain, we can still explain intentional acts by citing beliefs and aims that make them rational.

4.3.3

Another objection is that the Aristotelian theory confuses the concepts "person" and "human being." "Person" is a psychological concept, which applies to any being that is self-conscious, aware of its identity through time, and aware of other persons as persons. "Human" is a biological concept, whose ground of application is simply having a certain physical form. The human essence can therefore contain only physical and no psychological properties.

The Aristotelian theory does not confuse these two concepts but makes a substantive claim about them, one that denies that "human" is just a physical concept. The theory does not quite claim that humans are essentially persons: A being can have sophisticated rationality but not develop it in the specific way or to quite the degree required for personhood. The theory does claim, however, that humans essentially have the mental properties that, when developed, allow for personhood.[6] And it supports this claim with substantive arguments about thought experiments and rational psychology. The theory treats "person" and "human" as distinct concepts—there could, for example, be non-human persons—and links them, so far as it does, only on substantive grounds.

4.3.4

A final objection points to certain consequences of the Aristotelian theory. If humans are essentially rational, then foetuses and the irreversibly comatose are not humans, and babies and the severely mentally disabled may not be humans either. Are these consequences not implausible? And may they not have further implications, for example, for our treatment of the mentally disabled, that are morally repugnant?

One response to this objection is to revise the theory so it does not have these consequences. A revised Aristotelian theory says that what is essential to humans is not the actual presence of rationality, but the potential for rationality. Then a foetus is a human because, although not now rational, it will or may be rational in the future. Even the most severely mentally disabled—call them the "demented"—are humans because they could have been rational if their development had not been impaired.

This revised theory can be tied to a more general claim about biological kinds. Because these kinds develop through time, unlike an element such as gold, their essence must in general consist, not in actual properties, but in the potential for development that culminates in such properties.

To be acceptable, the revised theory must meet two conditions: First, the potential rationality that it makes essential to humans must be a real property, which persists through time and involves more than just the fact that a being will be rational-at-some-time or could be rational-in-some-possible-world. These last properties, which are indexed to times or worlds, cannot play the role essential properties play in *grounding* identities across times and worlds. If a foetus is identical to a future adult, the explanation is partly that it shares the adult's essential properties. But this explanation cannot be given using properties that presuppose identity, as indexed properties do. Second, the revised theory must allow degrees of perfection. In one sense of "potential," the potential for rationality does not admit of degrees: One either has it or does not. The potential that the revised theory makes essential to humans must be one that some who exercise rationality develop to a higher degree than others, and all who exercise rationality develop to a higher degree than those, such as foetuses, who do not.

If the revised theory can meet these conditions, Aristotelian perfectionism can agree that foetuses, babies, and the demented are humans. But I am not certain that the concept of potential required by the theory is available and will therefore assume that it is not.[7] This leaves us with the original Aristotelian theory and an opposite response to the objection: to defend the theory's restrictive claims about which beings are humans. If essential properties must be actual properties, there are two arguments for endorsing these claims.

The first argument is intuitive. If foetuses, for example, are humans, then it is possible for a human to exist without any mental life, let alone one that involves sophisticated rationality. But what is possible for one human must also be possible for all humans. (If a property is not essential to humans, there is no reason why all humans could not lack it.) This implies that a possible world in which beings with our bodily form never attain a mental life could contain humans—which is highly counter-intuitive. We do not count beings who never have thoughts or aims as belonging to *Homo sapiens* (4.2.1).

This argument assumes that essential properties must be actual properties. If the revised Aristotelian theory was true, the fact that one human lacks a mental life would not show that all humans could lack a mental life. If this theory is rejected, however, admitting foetuses as humans has intuitively unacceptable consequences.

The second argument is explanatory. It is well known that the explanatory method can change our view about the boundaries of a kind. We see what properties explain paradigm members of the kind, and count as further members only beings that share those properties. Sometimes unexpected candidates pass this test, and sometimes likely ones fail. Heavy water turns out to be water; fool's gold is not gold. Foetuses, it turns out, fail the test of humanness. If rationality explains the behaviour of paradigm humans, and foetuses are not rational, they are not humans. Our initial classification of foetuses is overturned when we learn that they are not explained in a fully human way.

If denying that foetuses are humans seems odd, one reason may be an ambiguity in the word "human." In a narrow sense, this word applies only to full-fledged members of the human kind, that is, to human beings. In a broader sense, it applies to anything connected with the human species; thus, we speak of human hair, a human corpse, and human sperm. That foetuses are not human beings does not mean they are not human in the broader sense: they are human foetuses, not dog or chimpanzee foetuses. But their being human in this sense does not imply that they are members of the species.[8]

A second reason may be a metaphysical assumption: that every living thing belongs to some natural kind. Because a foetus does not belong to the dog or chimpanzee kind, it is argued, it must belong to the human kind. As a little reflection shows, however, the metaphysical assumption is false. A human sperm or egg is a living thing and may develop into a full-fledged human, but it is not on any view now a human. A newly fertilized egg is also a living thing but not now a human because it lacks a human's bodily structure. At the other end of life, there can be a living body descended from a human that is not a human. On the "brain-death" criterion, a body can have a working heart and lungs yet not be a human because its brain has ceased to function.[9] (The brain-death criterion says that an individual has died. But, if the individual was a human, a human has died.) As these examples remind us, it is not sufficient for a human to exist that the material of a human body be present. This material must be organized in the right way, instantiating the properties and performing the activities of a full-fledged human. And it is a substantive question what these properties are and when they are and are not present. It is in principle possible that they all appear at conception and persist until the last organ shuts down. However, our two arguments have shown otherwise. The properties essential to humans are present in their entirety only when rationality has emerged, and remain only as long as it does.

What, then, is the status of foetuses, babies, and the demented? Although foetuses are not humans, they are descended from humans and may later turn into humans. They are closer to the human species than to any other species and can therefore be classed as almost-humans. Babies are probably also almost-humans, at least for a short time after birth. (Soon after birth, babies start a complex process of experimenting with concepts for understanding the world. Even to begin this process they must have some sophisticated rationality, which suffices to make them humans.[10] The demented are likewise almost-humans, although many others of the mentally disabled are humans. Their intellectual powers may not equal those of other humans, but this limitation is not decisive. So long as they have some sophisticated rationality, and many of them do, they are full-fledged humans.

These points bear on questions about our treatment of these beings. On any view, what matters morally in a foetus or baby is that it develop its capacities in later life, and regardless of its present status, perfectionism can tell us to promote this development.[11] As for the demented, even if they are not humans, a generalized perfectionism (1.4) can say they have a partial nature, perhaps involving unsophisticated rationality, that it is good for them to develop. In fact, far from having repugnant consequences for our treatment of the mentally disabled, perfectionism makes very much the right claims. With Aristotelian values, we cannot be content to keep the mentally disabled alive or to give them passive pleasures in a controlled environment. We must help them exercise as far as possible whatever powers of self-regulation they have. The rationality that is good in our lives is good in theirs, and we must encourage their efforts, however restricted, at autonomy and self-direction.

4.4 The Wrong Explanations?

The most important objection to perfectionism remains the wrong-properties objection, but given our recent discussions, it must take a special form. To claim that human nature contains morally trivial properties, it must point to an explanatory theory in which these properties play a central role. There are several possibilities.

4.4.1

A common version of the objection cites the explanations of evolutionary biology. Humans emerged from natural selection, and there are theories that explain why they survived this process and other species did not. Surely, critics argue, the properties central to these theories should be essential to humans, and surely if they are, perfectionism has unattractive consequences. To succeed in evolutionary competition, a species must be aggressive and prepared to trample others underfoot. Is aggressiveness then a value to be maximized? More generally, the property central to all evolutionary explanations is that of being a gene transmitter. Only by replicating its genes for another generation does an organism figure in evolutionary explanation. But this property seems morally irrelevant. Why should the number of genes a human transmits have any bearing on the intrinsic value of his life?

If the properties central to evolutionary biology were essential to kinds, this would indeed be damaging to perfectionism. But there are good reasons to think they are not. These properties often do badly on the intuitive test. Many species' survival depends on factors such as their colouring, taste, or smell. Only by blending into their environment or repelling predators do they avoid elimination. But colouring and taste seem paradigmatically accidental properties, in that we can easily imagine the species's appearing different. This intuitive failing should not surprise us. There are three reasons why evolutionary explanations, for all their interest, are of the wrong kind to support claims about species' essences.

First, these explanations often cite the wrong kinds of properties. What explains a yellow butterfly's survival is not so much its being yellow as its being yellow against a yellow background or, best of all, its being yellow when its predators cannot detect yellow. But these properties, especially the last one, are relational. They are not confined to the species itself but involve a relation to other beings. For this reason, they cannot plausibly be part of the species's essence. As relational, the properties can be lost even though the species itself does not change. They can be lost because its environment alters or because its predators increase their powers of detection. Because evolutionary biology aims to explain why some species were selected *above others,* it attends specially to relational properties, and the relational properties it focusses on are not plausibly part of one species's essence.[12]

Second, evolutionary explanations explain the wrong kinds of facts to reveal essences. They do not use some of a species's present properties to explain its other present properties, but explain how there came to exist beings with all those properties in the first place. They explain a species's origin, not its present behaviour, and it is present behaviour that must be explained to identify essences. Let us grant that the human species emerged from natural selection. Qualitatively identical beings, with identical properties related in identical ways, could have originated differently, for example, by divine creation ex nihilo. If they had originated in this way, they would surely still have been humans. The same theories would have explained their present behaviour, with the same properties playing a central role. The *individuals* resulting from divine creation would have been different; thus, neither you nor I would have existed. This is irrelevant, however, because origins are essential to individuals, not to kinds (2.2.1). It is again helpful to think about gold. Gold originated in a particular way, through some particular natural process, but it could surely have originated differently, say, in a scientist's laboratory. If what appeared in the laboratory had the same atomic structure as gold, would anyone deny that it was gold? As this example underscores, the explanation that reveals a kind's essence is not of its origin but of its present properties. Evolutionary biology does indirectly explain present human properties. It explains how certain other explanations—physical and rational explanations—came to be available. But it does not itself provide these explanations. It does not itself explain our present properties and therefore does not bear on questions about our nature.

Finally, evolutionary biology explains at the wrong level to reveal essences. To identify human essential properties, we need a theory that explains the behaviour of individual humans. When applied to organisms (rather than to genes), however, evolutionary biology explains facts about whole species: It is a theory of group, not individual, origins. Because of this, its central properties need not be present throughout a kind it explains. A species's colouring may explain its survival even though occasional individuals are born albino, or aggressiveness preserve it though some members are timid. This is especially true of the property at the centre of evolutionary biology, that of transmitting genes. I have not transmitted any genes to date in my life, and I may never do so. But this fact will surely not stop my life from being a human one: I will not be excluded from the species if I fail to reproduce.[13] The human species will not survive unless *some* humans transmit genes, but it is not required that we all do so and certainly not required that we do so essentially. In its application to organisms, evolutionary biology operates at the level of groups. It cares only that properties be present somewhere in a species and cannot determine what all the species's members must possess.

Although evolutionary explanations are important, they fail in three ways to have the content needed to reveal essences. It is not surprising, therefore, that they often highlight properties that by intuitive criteria are accidental.

4.4.2

Are there explanatory theories other than evolutionary biology that could ground a wrong-properties objection? Microphysical theories are excluded by our concept of human nature (2.2.2, 4.1.1), but there are many theories in the social sciences, from global theories such as classical economics and functionalist sociology to specific psychologies such as Freudianism. May they not highlight trivial properties?

Beginning with the global theories, they seem either to confirm the Aristotelian theory or to fail, for one of the same reasons as evolutionary biology, to support claims about essential properties. On one of two possible views, economic and sociological explanations reduce to those of individual psychology. They apply and extend the conclusions of rational explanation and, by so doing, increase its reach. According to this view, obviously, the theories cannot support new claims about essences but instead strengthen the old ones. According to the rival view, economics and sociology use autonomous principles or refer to autonomous facts true at the level of groups. But then their central properties need only appear somewhere in the group and need not be had universally as essence requires. On the first view, the global theories only confirm the Aristotelian theory; on the second, they explain at the wrong level to reveal essences.

Specific psychologies such as Freudianism raise different issues. They add to the general structure of rational explanation some specific theses about what humans believe and intend. These theses tend not to do well on the intuitive test. Even if humans actually desire oral gratification or repress anxieties, we can imagine possible worlds where they do not. This failing again has a structural source. Theories like Freudianism are not self-verifying: if they seem plausible, it is because they offer to extend rational explanation or to make more human behaviour rationally explicable. Freudians claim that, by ascribing specific mental states, we can explain seemingly fortuitous acts as intentional and much apparently unconnected behaviour as revealing a single motive. Their theory's attraction, in other words, is that it promises more coherent explanations or explanations that do better by the principle of charity (4.2.3). The theory is therefore best seen as an application of rational psychology rather than as a rival to it. Taking the general idea of rational explanation, it claims that we will understand humans better and find more coherence in their behaviour if we ascribe to them certain mental states. In this claim the specific ascriptions are secondary, and a generic rationality, as before, is the crucial explanatory property.

4.4.3

The wrong-properties objection is hard to answer conclusively, for one never knows what someone will claim is essential to humans. But my general argument has been this. If essences are discovered through explanations, the claim that humans have a trivial essential property must point to a serious science where this property plays a central role. Critics often mention evolutionary biology, but it is in several ways of the wrong kind to support claims about essences. When we ask for other theories that do not govern inanimate matter, we find only explanation by organic structure and rational explanation as serious candidates. Unless there are sciences I have missed, this leaves the Aristotelian theory of human nature and the promising Aristotelian perfectionism it supports. Starting from the general perfectionist idea, and identifying nature with living essence, we have arrived at a perfectionism with three intrinsic goods: physical perfection, which develops our physical nature, and theoretical and practical perfection, which develop theoretical and practical rationality.

II

ARISTOTELIAN PERFECTIONISM

5

The Basic Structure

To develop perfectionism further, we must know more about its *structure* and its *content*. We must know what formal principles it uses to arrive at judgements about particular acts and what specific human states best instantiate its values. How does the theory move from the Aristotelian ideal to particular moral judgements? And how does it flesh out that ideal?

Part II addresses these questions, moving in each case from general issues to ones that are more concrete. This chapter defends some general claims about moral structure. It argues that the best perfectionism is a maximizing consequentialism that is time- and agent-neutral, telling us to care equally about the perfection of all humans at all times. Our ultimate moral goal, according to this perfectionism, is the greatest development of human nature by all humans everywhere.

Before we begin, two points bear repeating. One is that a study limited to pure perfectionism does not imply that this morality is the one everyone should accept (see section 1.4). Perhaps perfectionist ideas will serve best in a pluralist morality, where they are weighed against claims about utility or rights. Even so, exploring perfectionism on its own is the best way to discover its importance and the contribution it can make to more inclusive views.

The second point concerns broad and narrow perfectionisms (1.2). The theory we will discuss has foundations in human nature but can be embraced apart from these foundations. Someone skeptical about essences may find a merely broad perfectionism containing theoretical, practical, and physical values appealing and may adopt it without tying it to any claims about human nature. This will sever his theory from an ideal many find compelling and also diminish its unity. But it will leave a plausible morality, one that readers unconvinced by my part I can still consider seriously.

5.1 Maximizing Consequentialism

Perfectionism clearly directs us to develop our human nature. But how far, and on what basis?

5.1.1

Historically, perfectionism has always been a *maximizing* morality, which tells each human to (help) achieve the greatest perfection he can. It has wanted its ideal promoted not just to some degree, but as far as possible. Aristotle, for example, thinks there is a better part of the soul and wants us to "strain every nerve" to develop it. Unlike instrumental goods, he argues, whose usefulness has an upper limit, intrinsic goods are always preferable in greater quantities.[1] Wilhelm von Humboldt's goal for a human is, not some development, but "the highest and most harmonious development of his powers to a complete and consistent whole,"[2] and Bradley says the best individual "most fully and energetically realizes human nature." To be a good human "in all things and everywhere, to try to do always the best, and to do one's best in it, . . . this and nothing short of this is the dictate of morality."[3]

Maximizing is not the only possible structure, for perfectionism could take a *satisficing* form. It could care that humans develop their nature to some reasonable degree but be indifferent about what they do beyond that. Then agents would be required to strive up to a threshold of perfection but free to continue or not once the threshold was reached.

Satisficing may be attractive for some values, for example, pleasure or desire-fulfilment,[4] but not for an ideal of perfection. If we are attracted by this ideal, it is as something to be maximized and pursued to the highest degree. This is reflected in our intuitive judgements. Imagine that Mozart's musical talents are such that, with only a small effort, he can reach the satisficing threshold for musical achievement and for achievement generally. (If this is not so, the threshold is set so high that it will never come into play for most people.) Then a satisficing perfectionism says that Mozart has no duty to expend more than this small effort, which is highly counterintuitive. Intuitively, the perfectionist duty to develop one's talents, considered just as a self-regarding duty, applies no less to those with greater talents. As their achievements increase, the demand to build on them does not diminish. If anything, we can understand the view that the duty to develop one's talents is more pressing for those with greater talents (see further 6.2.1). Satisficing's ill fit with perfectionist values is reflected in our language for these values. We would not say of someone who was content with a reasonable development of his talents that he aimed at "excellence" or was dedicated to "perfecting" himself. As our theory's name indicates, a concern for human development goes naturally with a maximizing approach.[5]

It may be objected that we cannot really distinguish maximizing and satisficing, given the imprecision of any plausible ideal of excellence. We cannot measure human development on a precise cardinal scale and hence cannot distinguish reasonably good human states from those that are best.

Measures of perfection are indeed partly indeterminate (see chapters 7–10), but not entirely so. Consider T. S. Eliot. Early in his life, he had to choose between returning to the United States to teach philosophy and continuing his literary career in England. Surely, if he had returned to teach philosophy, he would have led a reasonably good life. Surely also, by remaining in England and writing poetry, he led a better life, with greater achievements. If so, maximizing and satisficing make different judgements about Eliot's choice. Satisficing says it would have been permissible for Eliot to teach philosophy; maximizing says that this would have been wrong and a waste of his finest gifts. We cannot assign precise measures to the values in either of Eliot's possible lives. Even so, we can be confident that, although both would have been above a threshold, one would have been further above it than the other.

The discrimination possible here between life options is even more common within them. Once we have chosen an activity such as philosophy or poetry, we can often distinguish quite finely between achievements within it. We can distinguish between a good philosophy paper and a slightly better one, or between a creditable line of verse and one that revision has subtly enhanced. In fact, it is within activities, once these are somehow chosen over others, that maximizing has its clearest implications and also its greatest appeal. On a satisficing view, our philosophers and poets could content themselves with works that are reasonably good. Given maximizing, they must strive to make their philosophy as incisive and their verse as fine as possible.

There are other alternative structures than satisficing, but they fall to a similar argument. One could combine the perfectionist ideal with a *respect* structure, in which a duty to promote the ideal is constrained by a stronger duty not to choose directly against any of its elements.[6] But, whatever its fit with other values, this structure does not fit an ideal of development. It implies that it is always wrong to destroy some archaeological evidence, and thus the possibility of some knowledge, to reach more important evidence below, and wrong to intentionally close oneself off from one good activity to preserve oneself for another (think of a pianist ensuring that he will never try boxing). These claims are counter-intuitive because perfectionist values come to us naturally as ones to be promoted, and more specifically, to be maximized.

5.1.2

That perfectionism should have a maximizing structure does not imply that it should be *consequentialist*. Consequentialist theories, which can be maximizing or satisficing, are distinguished by the way they relate the concepts "good" and "right." They treat "good" as explanatorily prior and always identify the right act by how much good it produces. A consequentialist perfectionism says that human perfection is good and that agents ought to maximize it because it is good. A non-consequentialist perfectionism says only that agents ought to maximize perfection.

This account of consequentialism assumes that "perfection is good" says more than "perfection ought to be maximized," and that the first statement can explain the second. These assumptions need defence.

Whatever its explanatory role, the concept "good" allows more evaluations than are possible using just the concept "right act." Imagine that a moral theory tells agents to maximize a state *s*, and that *A* and *B* are trying to do so. *A* has available to her acts that will produce 10 and 100 units of *s*, while *B* can produce either 100 or 1,000 units. *A* chooses the act that produces 100 units of *s*, and *B* produces 1,000. With only the concept of right action, the theory cannot make different evaluations of these agents' acts. *A* has done both what is right and what she ought, and so has *B*. With the concept "good," however, the theory can discriminate. It can say that *B* has produced more good than *A* or brought about better consequences. Something similar is possible in cases involving risk. Imagine that *A* and *B* both choose an act with a high probability of producing 1,000 units of *s* and a low probability of producing 10 units, but while *B* does produce 1,000 units, *A* is unlucky and produces only 10. If the concept of right action is tied to probabilities at the time of acting, a theory using only this concept cannot evaluate *A*'s and *B*'s acts differently.[7] A consequentialist theory, however, can say that *B* has produced more good.

As these examples show, "good" allows evaluations that depend on factors outside people's choices. These factors may precede a choice, and determine the options from which it is made, or follow and determine its effects.

We can account for these judgements if we treat "good" as a primitive concept, but then it is unclear why it should be explanatory. And there is another possibility: to define "good" in terms of "ought," but "ought to desire," not "ought to do." Because desire is not constrained by circumstances, as action is, the scope of "ought to desire" is broader than that of "ought to do." In the first example, in which *A* can produce only 100 units of *s*, he can wish that he were able to produce 1,000, as *B* is, and *B* can be glad to have more fecund options. In the second example, *A* and *B* can both hope their acts will produce 1,000 units, and be, respectively, chagrined and pleased by what occurs. The distinctive evaluations "good" permits are captured by a definition of the good as what agents ought morally to desire.

And this definition explains why "good" is explanatory. Alongside the categorical imperatives of morality is a general hypothetical imperative, or imperative of rationality, telling agents to avoid situations where they desire an end but do not take the most effective means to it. Given this hypothetical imperative, any moral command to desire a state entails further moral commands to perform the acts that bring it about. Given general canons of rationality, in other words, categorical requirements on desire entail and so explain further categorical requirements on action.

I will interpret consequentialism as using this definition of "good." A consequentialist perfectionism tells us first to desire the state in which human nature is developed to the highest degree and then, assuming we are rational, to promote it.

5.1.3

Given this account of consequentialism, why think it a desirable structure? Why want perfectionism to start from judgements about "good"? Except in special circumstances, non-consequentialist maximizing theories are not very plausible. They tell agents to promote some state—on our assumptions, to maximize it—but let them be indifferent to the amount they actually produce. Although agents must search diligently for the act most productive of *s*, for example, they need not care if little *s* results. The dislocation of attitudes here is disturbing. Agents are allowed to have acts and desires that do not fit together in the normal way. If a moral theory addresses the person, it ought surely to address the whole person: not just his acts, but also the desires and attitudes behind them.

Most perfectionists avoid this dislocation by making judgements that are not just about right acts. Aristotle, for example, thinks a person's perfection depends in part on her choices and will be reflected in her response to difficult trials.[8] But he also thinks it depends on factors preceding her choices. These include instrumental goods such as wealth, friends, and political power, without which the best life is not possible.[9] They also include her native abilities.[10] Like the vast majority of perfectionists, Aristotle holds that someone who achieves more perfection only because she was born with vaster abilities nonetheless counts as better. Imagine that one person with the talent for as many as 100 units of perfection makes a moderate effort and achieves 75, while another capable of 50 units exerts herself tremendously and achieves 49. Aristotle would say that the first is a better human, and her life is a better life, even though the second's choices are more commendable. Aristotle also gives weight to factors following choice. He does not think bad luck in the outcome of a choice can have a large effect on a person's excellence or make an otherwise good life bad, but he does think it can prevent her from reaching the highest perfection.[11]

Aristotle's attraction to these judgements is tempered by his belief that perfection should be self-sufficient, so each person can achieve it on her own.[12] If perfection were entirely self-sufficient, there would be little scope for distinctively consequentialist judgements. Each person would always have available to her an act certain to produce her excellence, and her failure to achieve excellence could always be attributed to wrong choosing. Aristotle does not believe that perfection is entirely self-sufficient, but Kant does. Kant equates perfection with a morally good will and says that anyone can acquire this will from one moment to the next. Faced with wrongdoing, we can always regard it "as though the individual had fallen into it directly from a state of innocence."[13] For Kant a person's perfection depends entirely on her choices, which makes much of the point of consequentialism disappear. More generally, whenever the goal in a maximizing theory is one agents can achieve on their own, a consequentialist formulation adds little that is new and non-consequentialism is plausible. But most perfectionists do not follow Kant's line. They characterize human nature so that its development, although depending in part on a person's choices, depends also on factors outside them. And this dependence makes a consequentialist structure overwhelmingly preferable.

5.1.4

Despite these arguments, some philosophers deny that perfectionism is consequentialist. This denial may be because the theory lacks a common mark of consequentialism.

It is well known that consequentialism cannot be defined as the view that evaluates acts just by their consequences because what belongs to an act's consequences depends on how it is described. An act of "pulling the trigger" may have someone's death as an external consequence; the same act described as "killing" does not. Nonetheless, it is true of many familiar consequentialisms that, relative to the descriptions we use in everyday life, they do evaluate largely by results. Hedonistic utilitarianism, for example, values an introspectible state of pleasure, and, given its passive nature, this state is usually external to the (conventionally described) acts that produce it.

In perfectionism, by contrast, the good is largely active. To achieve it, a person must do things and pursue plans that engage with, and have an impact on, the world (chapters 8–10). Because of this, her acts, even as conventionally described, cannot always have perfection as an external consequence. They must sometimes embody perfection, and contribute to the good by instantiating it now rather than by allowing its occurrence later. In the language of commentators on Aristotle, they must sometimes be "constitutive," not "productive," means to excellence,[14] or components of, not preconditions for, the good. This does not prevent them from being, in a broad sense, means, nor does it stop the theory that requires them from being consequentialist. This theory starts from a vision of the good and always commends acts to the degree that they promote the good. Still, that perfection is active can make the theory's consequentialism harder to see.

5.2 Time- and Agent-Neutrality

If the best perfectionism starts by identifying some human states as intrinsically good, the next question is how it does so at different times and for different agents.

5.2.1

Consider first the theory's directives about one life. Do they give a person the same moral goal at all the times in her life, so all her acts can converge on one good, or do they give her different goals at different times? On the first or *time-neutral* view, the theory has an overarching ideal of the best human life, and what a person should seek at every time is to do what most contributes to this life. On the rival *time-relative* view, there is no such overarching ideal, and a person's aim at each time may be just to maximize her present development of essential properties.

A parallel issue arises across persons. Perfectionism can assign the same moral goal to all persons, one involving (in some sense) the greatest perfection of them all, or it can assign different goals to different persons. In an *agent-neutral* perfectionism, there is an overarching goal for all humanity, which all can pursue together. With *agent-relativity* there are individual goals for individual agents, and the possibility of conflict among them.[15]

These issues are separate from ones about consequentialism. A theory can be time- and agent-neutral but not consequentialist, making no claims about "good," and it can also be consequentialist and relativized. In the second case its judgements of value use not the simple concept "good" but the relativized concept "good from the point of view of person x at time y." This relativized concept is the fundamental one, in terms of which the others are defined: To call something simply "good from a person's point of view" is to call it "good from his point of view at all the times in his life," and to call it "good period" is to call it "good from the point of view of all persons always." The concept is also definable in terms of desire. If some state is good "from the point of view of person x," then x (and perhaps only x) ought to desire it, whereas, if it is good generally, everyone should desire it. If a moral theory starts by specifying an object of morally required desire, this object can be the same for all persons and times or it can be different. Which structure yields the best perfectionism?

5.2.2

A plausible perfectionism must clearly be time-neutral. The idea that humans should maximize just their present development of essential properties, for example, has no intuitive appeal and has never been held. On the contrary, perfectionists repeatedly make claims about "the best human life" (2.2.2). They say that excellence requires "a complete life," and in comparing the goods of theory and practice weigh, not knowledge and action as such, but whole lives devoted to them, for example, the philosophical and political lives.[16] This is no mere stylistic point. The persistent talk of "the good life" implies that we should have a constant goal throughout our lives, one composed in some way of states at all the times we live.

Time-neutrality is not appropriate for all consequentialist theories; it does not, for example, suit theories that value pleasure and the absence of pain. Imagine that, awaking in hospital with short-term amnesia, you are told that you are either a patient who had a very painful operation yesterday or a different patient who will have a less painful operation today.[17] As you await further news, which should you hope you are? A time-neutral hedonism says you should hope you are the second patient, who has less pain in his life as a whole. But in your position most people would hope fervently that they were the first patient, whose pain is over. They would prefer more rather than less pain in their lives, so long as it was in their past. This attitude is too natural to be condemned. A plausible hedonist theory must accommodate it by giving less or no weight to past pleasures and pains. At each time, it must tell agents to desire mainly or only their satisfaction from then into the future.

This departure from neutrality is not needed in perfectionism. Imagine that, awaking in hospital with temporary amnesia, you are told that you are either a scientist who made a major discovery last year or a different scientist who will make a minor discovery next year. You will surely hope that you are the first scientist. You will want your life to contain the greatest scientific achievement possible, regardless of its temporal location. Unlike past pleasures and pains, past perfections do matter to us today. Our past achievements can still give us feelings of pride; past failures can make us cringe. This is captured in a fully time-neutral theory, which tells us to desire past perfections as much as future ones and even, given the right account of perfection, to actively promote past perfections (see 8.4.3).

It is consistent with time-neutrality to give more weight to some times than to others. A theory can say that perfections in one's thirties, for example, count more than perfections in other decades, so long as at all times one should care about one's thirties more. I assume, however, that the best perfectionism does not follow this line. It sees all times as equally parts of a life and therefore values them equally. (One's perfections in one decade may be greater than in others and therefore count for more. But this is a matter of the perfections' magnitude, not the time at which they occur.) The theory's ideal is composed of states at all the times in a life, and composed in a way that treats times equally. Perfectionism is not just neutral with respect to times but also impartial about them.

5.2.3

Neutrality across agents is more controversial, especially because on any view perfectionism gives a central place to self-regarding duties. It tells each person to seek his own perfection, or to develop his talents, thus finding an important duty where many moralities find none (1.3). Partly for this reason, many philosophers assume that perfectionism must be purely agent-relative, telling each human to develop just his own nature. In fact, the issue is open. Should perfectionism contain only the one, self-regarding, duty, or should it extend its goal so that each human must care also about the perfection of others? More specifically, should it require each human to care equally about the perfection of all?

It can be argued that this extension is a requirement of consistency. If perfectionism is impartially time-neutral, giving equal weight to all the times in a life, a certain symmetry may require that it also be impartially agent-neutral. By abstracting from temporal particularity, the theory may commit itself to abstracting, in a parallel way, from particularities about persons.[18]

This argument may not be decisive, but it is suggestive. And even without it, there is an overwhelming case for agent-neutrality. In defining human nature, we rejected moralism, the view that developing one's own nature requires other-regarding virtues. Although humans are essentially rational, their rationality can be realized as much in skilful burglary as in philanthropy (2.3.2). Without moralism, a fully agent-relative perfectionism can permit and even require acts that diminish others' perfection, as long as they advance the agent's. This is morally unacceptable. Our project is to develop, not some perfectionism, but the perfectionism that accords best with our considered judgements (3.3.1), and central among these judgements are ones forbidding harm to others' good. Aristotelian perfectionism cannot capture these convictions if it is fully agent-relative, but it can if it is agent-neutral. Then each agent's duty to develop his own rationality is constrained by an equal duty to preserve and promote rationality in others. If an act that harms others does more to set back their perfection than it does to advance the agent's—as it normally will—the act is wrong. The benefits to the agent are outweighed by losses to his victims. In the same way, if increasing one's own good involves failing to make a larger contribution to others' good, that too is wrong. An agent-neutral structure is not sufficient for an acceptable perfectionism; the theory must also have plausible content. But something like it is at least necessary. If a non-moralistic perfectionism is to match our moral convictions, it must supplement the duty to pursue one's own perfection with a duty to preserve and promote others'.

This argument does not decisively support full agent-neutrality. Perfectionism could capture some other-regarding duties if it told agents to have some concern for others' perfection but a greater concern for their own, and this mixed structure—an agent-neutral duty but also an independent agent-relative one—may be understandable given perfectionist values.[19] But perfectionism surely has the best consequences—those that best fit our particular judgements—if it tells agents to care equally about the perfection of all.

If it is fully or partly agent-neutral, perfectionism must distinguish sharply between questions about what makes a person's own life best and questions about how she ought, all things considered, to act. This distinction is familiar from other consequentialist theories. Utilitarians think the best life is the most satisfying, but they think it often right to sacrifice one's own satisfaction to benefit others. The same distinction is needed in perfectionism. Its best version still says that the best life is the most rational, but it adds that we must sometimes limit our own exercise of rationality to preserve or promote others'. This last claim yields the right conclusion—that it is wrong to harm others—and also gives it the right rationale. A moralistic agent-relative perfectionism can forbid acts that diminish the perfection of others, but its ultimate reason for forbidding these acts is that they make the agent's life worse. This is not the right ultimate reason. If harming others is wrong it is so most fundamentally for what it does to *them*, for how it affects *their* lives. Only as agent-neutral can perfectionism give this explanation.

Although it captures other-regarding duties, agent-neutral perfectionism does so in a distinctive way. In the moralities most studied by philosophers, we are to help satisfy others' desires, whatever their content, or respect any self-regarding use of their freedom. This fits the idea that there is nothing those people should seek within their lives, and it cannot survive if that idea does not. If there are better and worse ways others can live, our duty must be to help them live better. It must be to help develop in them what is most worth developing. In many cases, this has familiar consequences: We are not to kill others or maim or torment them. In other cases it does not. If a person wants to waste her talents or to live in idleness, she does not have the same claim on our aid as if she sought something worthwhile. Even in the familiar cases, our duty has a distinctive ground: The reason it is wrong to kill or maim others is that this will prevent them from achieving full human development. In agent-neutral perfectionism we are to value others, but in a distinctive way: as potential achievers of a human ideal.

There is a final, more abstract argument for agent-neutrality: In choosing this structure, we remain truest to the original perfectionist idea and to our original valuing of human nature. Return to the self-regarding case and imagine that I seek knowledge in myself because I think knowledge develops human nature. Is my thought that my knowledge is good only from my point of view, entailing nothing about how others should view my achievement or how I should view theirs, or is my thought impersonal? Surely in the normal case it is impersonal. In the normal case I view my knowledge as an instance of something that is (simply) good, with value wherever it appears and a claim to recognition by anyone aware of its presence. But then even my self-regarding thoughts are implicitly agent-neutral and take me beyond myself to value perfection in all humans. They treat my knowledge as good not from one but from all points of view, and so imply judgements about the value of knowledge in other people.[20] There is no logical requirement that perfectionism be agent-neutral; its basic idea could be read as agent-relative. But I do not think this would retain the idea's full intuitive appeal. At its most compelling, the idea is that human development is (simply) good, and (simple) goodness is goodness of an agent-neutral kind.

5.2.4

There are, then, three arguments for agent-neutrality: It mirrors perfectionism's approach to times, promises the most appealing consequences, and is truest to the original perfectionist idea. Still, many philosophers assume that perfectionism must tell each human to develop just his own nature.

In part, this assumption is encouraged by the perfectionist tradition. When writers such as Bradley refer to human development as "self-realization," they invite the thought that each is to care only about himself. And there are more pervasive factors. Many perfectionists accept strong natural desire doctrines (3.1.2), on which they build naturalistic justifications of perfectionism (3.2.1). But the desire doctrines are always agent-relative. They say that humans desire, not everyone's perfection, but just their own. Because the only moralities that could follow from such doctrines are themselves agent-relative, we are again encouraged to ignore neutrality. Finally, many classical theories are moralistic, which makes the issue of neutrality less pressing. If achieving one's own perfection already involves treating others well, there is less need for a further demand to promote their perfection. I have argued that agent-relative perfectionism cannot give the right explanation of other-regarding duties, one that makes them truly other-regarding (5.2.3). But if it at least contains these duties, there is less call for worry about its structure.

Nor is it only the tradition that encourages an assumption of agent-relativity. Even the best perfectionism, free of moralism and of any strong desire doctrine, has features that can obscure its agent-neutrality. Certain empirical facts lead it, when applied deliberatively, to emphasize the duty to seek one's own perfection over the duty to seek others'. These facts deserve examination.

5.3 The Asymmetry

5.3.1

Given utilitarian values, an agent-neutral structure leads to agent-neutral moral thinking. People can make their own lives pleasant or satisfying, and they can do the same for others by giving them money, admiration, or good food. In deliberating, therefore, they should give roughly equal weight to all people's good.

In Aristotelian perfectionism, by contrast, there is an *asymmetry* in agents' ability to bring about the good, one that makes them less able to promote others' perfection than their own. In favourable conditions they can produce their own excellence directly, but they have less power over others'.

One reason for this asymmetry is that much Aristotelian perfection is active (5.1.4). Whereas pleasure is a passive state, both physical and practical perfection involve doing things, forming goals and realizing them in the world. And each person's doing must be largely her own, reflecting her own energy and commitment. Others can avoid interfering with her activity, and they can offer encouragement and needed resources. Yet, they can rarely produce her perfection themselves: Past a point, her achievement of active goods must be her own.

A second reason is that the rational perfections involve complex inner states. For theoretical perfection a person must have knowledge, or justified true beliefs in an organized explanatory structure (chapters 8–10). And, as "Socratic" theorists of education insist, this knowledge can develop only internally. Others may give elementary instruction, but true understanding, or a systematic insight into phenomena, must be a person's own achievement. In the same way, practical perfection involves acting from complex, structured intentions. A person's goals must be sophisticated in their scope and inter-relations (chapters 9–10), which must likewise be her doing. When perfection is not only active but inner, it is even more a good that each must largely achieve on her own.

The asymmetry these facts support is consistent with agent-neutrality but can also obscure it. If people can achieve more in their own lives, they should direct more of their energy there. In principle they should care equally about everyone's perfection; in practice, they should attend disproportionately to their own.

5.3.2

The perfectionist most impressed by this asymmetry is Kant. He says we should desire the perfection of all rational agents but can act only to promote our own:

> It is contradictory to say that I make another person's perfection my end and consider myself obligated to promote this. For the perfection of another man, as a person, consists precisely in his *own* power to adopt his end in accordance with his own conception of duty, and it is self-contradictory to demand that I do (make it my duty to do) what only the other person can do.[21]

This extreme claim depends on Kant's narrow equation of perfection with the morally good will, and even then it may be overdrawn if moral education is possible.[22] But Aristotelian perfectionism cannot affirm anything like Kant's absolute asymmetry. It supports only a more moderate claim.

The asymmetry's central claim is that, when perfection is active and inner, we cannot provide causally sufficient conditions for it in others. But it does not deny that we can provide causally necessary conditions or conditions that make another's excellence more likely. Negatively, we can refrain from killing another or disrupting her projects. Positively, we can sometimes give aid or supply resources. A plausible asymmetry must recognize these possibilities. In particular, it must recognize that we can often provide necessary conditions for others' good and have a moral duty to do so. We may even have a strong duty to share resources with those whose development is hampered by poverty. But these qualifications still allow a substantive causal claim. In our own case, when the necessary conditions for perfection are present, we can normally provide sufficient conditions; for others we cannot.

In a maximizing theory, the distinction between necessary and sufficient conditions is not of intrinsic moral significance. What matters in such a theory is how much good an act promotes, not how it does so. But the distinction can be important in practical deliberation. If I have a choice between certainly achieving my own perfection and perhaps contributing to another's, I should often prefer the former, and when I do help others, it can often be without conscious reflection. This is clearest for negative contributions such as not killing and not maiming. I can develop a general disposition to avoid these harmful acts, and an agent-neutral perfectionism says I should. It tells me to internalize a rule against diminishing others' perfection and to follow it habitually. Then my conscious reasonings can concern my positive acts, and these, given the asymmetry, will often promote my own good.

This point is strengthened if my society has laws forbidding acts that harm others. Then I can provide many necessary conditions for others' good by simply obeying the laws, with no thoughts beyond myself. The same holds, if less simply, for the sharing of material resources. A well-run society has economic rules that, if adhered to by all, distribute resources in the way most conducive to everyone's perfection. Operating within these rules, agents can seek resources for themselves, confident that in so doing they are not depriving others who have a greater perfectionist claim. In fact, perfectionism favours a general division of labour between governments and private citizens. Governments establish rules—legal, economic, social—whose proper functioning promotes the perfection of all. Given these rules, citizens are then free to concentrate on their own good. So long as they obey the social norms protecting others, they can aim largely at perfection in themselves.

This division of labour is endorsed by Aristotle. At the deepest level, Aristotle's perfectionism is not just agent-relative. The *Nicomachean Ethics* says that, although the ends of the state and the citizen are the same, "that of the state seems at all events greater and more complete whether to attain or to preserve."[23] And the *Politics* repeatedly identifies the end of the state with the good of all, and the ruler's practical wisdom with his ability to legislate for the good of all.[24] Why then is there so little reference to others' good in Aristotle's ethical writings? Why does the citizen's practical wisdom deal so exclusively with his own *eudaimonia* and his own perfection? Aristotle's moralism is relevant here, but so too is the asymmetry. Aristotle believes that a good state has laws forbidding citizens to interfere with each other's pursuit of excellence. If he doubts that citizens can do much of a positive nature to help each other, he will think their other-regarding duties are largely exhausted by the duty to obey the laws, which does seem to be his view. Justice in the sense of obedience to law is the one virtue Aristotle discusses in terms of its effects on others, calling it "another's good,"[25] and relating it through the purpose of the laws to the *eudaimonia* or perfection of all.[26] This division of labour is not unique to Aristotle or to perfectionist theories generally, but these theories make it especially appropriate. By limiting so severely what citizens can do to promote each other's good, they make it especially plausible that agent-neutrality is handled at the level of politics.

5.4 Competition and Co-operation

5.4.1

If the asymmetry can obscure agent-neutrality, its effect is reinforced by another empirical fact about perfection, that it is not very *competitive*. It is not normally a condition for one person's achieving excellence that others not do so, nor do his attainments often exclude theirs. For the most part, perfections in different people are compossible. This fact again marks a contrast with utilitarianism. People often want things that cannot be shared—to be sole owner of this or winner at that—and pleasing one often means frustrating another. But why should one person's knowing a scientific law prevent others' knowing it? Why cannot different people exercise their bodies or pursue challenging projects simultaneously? That perfection is not competitive does not mean that agent-neutrality is unimportant, for there are ways of seeking one's own perfection that harm others'. The present point is that these ways are not usually intrinsically far preferable to others. To avoid them, one need only choose something equally or almost equally good in oneself. This, again, can make agent-neutrality less evident. If we associate neutrality with a demand to sacrifice one's own good for others, we may less readily see it in a morality where the demand is less frequent.

Non-competition is a central idea in Green's *Prolegomena,* where it serves as a condition on acceptable moralities. The distinction "between considerate Benevolence and reasonable Self-love," Green writes, ". . . is a fiction of philosophers." A true notion of good

> implies interest in an object which is common to all men in the proper sense,—in the sense, namely, that there can be no competition for its attainment between man and man; and the only interest that satisfies this condition is the interest, under some form or other, in the perfecting of man or the realization of the powers of the human soul.[27]

This claim of Green's is disputed by Sidgwick. Sidgwick notes that Green's ideal involves not just moral virtue but also achievements in the arts and sciences, and argues that "so long as the material conditions of human existence remain at all the same as they are now," the achievements of different people can conflict.[28] Here Sidgwick is clearly right. Like Kant's absolute asymmetry, Green's absolute claim of non-competition cannot be sustained.[29] But a weaker claim may still be true and important. Although Sidgwick's problem of scarce material resources is but one source of conflict in utilitarianism, it seems the only serious source in perfectionism. The wealthy can still have conflicting desires, but if we had all the resources we need, how often would my knowledge or exercise of skill diminish yours? If there was no competition for material resources, what further competition could there be? Even material scarcity may be less of a problem in perfectionism than in utilitarianism. On the Aristotelian account, perfection does not require great riches. It requires security, yes, and leisure, but past a point wealth loses importance and can even distract one from the good life (12.4). If so, even moderate abundance could remove the main source of perfectionist competition. Everyone would have the material means for a valuable life, and serious conflict would disappear.

5.4.2

Although the asymmetry and non-competition have the same initial effect, they are somewhat in tension, and there are two different ways of bringing them together. The first emphasizes and extends the asymmetry. It says that if people have too much to do with each other they will get in each other's way, and should therefore all seek excellence in isolation. This view appears in a mild form in Aristotle's defence of private property:

> Property should be in a certain sense common but, as a general rule, private; for, when everyone has a distinct interest, men will not complain of one another, and they will make more progress, because everyone will be attending to his own business.[30]

The view is most associated with Nietzsche, who repeatedly claims that solitude is necessary for true achievement. In Nietzsche's words, *Thus Spoke Zarathustra* is "a dithyramb on solitude,"[31] while another work advises, "Live in seclusion so that you *can* live for yourself."[32]

The opposed view extends the idea of non-competition, claiming that the perfections of different people *co-operate.* One person's excellence is often a necessary condition for others' or makes it more likely, and in turn is encouraged by theirs. This co-operative view appears in Green, whose "common good" is "a state of mind or character of which the attainment, or approach to attainment, by each is itself a contribution to its attainment by every one else."[33] It is also implicit in Marx's claim that under communism "the free development of each is a condition for the free development of all."[34] Its implication is the opposite of that endorsed by Nietzsche. Instead of isolating themselves, humans should as far as possible work together and provide through joint action some necessary (if not sufficient) conditions for each other's good.

It would be wrong to adopt either of these views to the exclusion of the other. Sometimes perfection requires isolation, sometimes it goes best in a group. But the second, co-operative, view has a certain affinity for Aristotelian values. Successful intellectual work is often communal, and the same holds for many practical pursuits. Games such as chess allow two people to exercise skill together, with the good play of one raising the level of the other's. What is more, a specific element in practical perfection is realized whenever people knowingly collaborate (10.2). It would be foolish to claim that perfections always co-operate, for they do not. But people often do best seeking excellence together. When this is so, agent-neutrality is even further obscured. The acts best for others are also best for oneself, and each can choose rightly by agent-neutral standards, given only agent-relative aims.

5.4.3

This chapter has argued that the best perfectionism is a maximizing consequentialism that is time- and agent-neutral, telling us to promote as far as possible the perfection of all humans at all times. These claims do not represent the only possible interpretation of the original perfectionist idea, but they are, I believe, the most attractive interpretation. Perfectionism could be developed as a satisficing or agent-relative morality, but the results would in each case be counter-intuitive. My proposed structure promises the most defensible moral theory and the one closest to what most perfectionists have held.

6

Aggregation

To say the best perfectionism is time- and agent-neutral is not yet to specify the state it finds intrinsically good. Even if we know what it is for one human to develop her nature at one time, we have not been told what combination of these states we are to maximize, either in her life or globally. We do not know what "the greatest perfection of all humans" amounts to, for we do not know how to *aggregate* perfections across times and persons.

Consider Achilles in *The Iliad*. He is fated to lead either a short, glorious life or a longer life of lower average achievement. To know which of these lives is better, it is not enough to know about their achievements at particular times; we must also know whether the greater number of achievements in the longer life makes up for their lower quality, and to know this we must have a principle for aggregating achievements across times. There are several candidate principles, each giving equal weight to all times; which principle is best? A similar issue arises across persons. We may know all the effects of two social policies on the perfections of individuals, but to evaluate the policies we must know how those individuals' perfections combine into an aggregate social perfection. Again there are several principles available, each giving equal weight to all lives; which is the best?

This chapter addresses these issues, using a simplifying assumption: that perfectionism has a precise cardinal measure of each human's development of his nature at each time. This assumption is unrealistic (5.1.1), but it will be analytically helpful (see 7.1.3). By asking which aggregative principles would be most attractive *if* perfectionism had precise cardinal measures, we can deepen our understanding of the less precise judgements it actually makes.

We will also assume that all relevant values are captured in these measures of perfection at a time.[1] Values that may seem to resist this treatment, such as the completion of a long-term project, can be handled if the value of a state at one time depends on facts about related states at other times (8.4.3, 9.2.3).

Finally, we will assume that perfectionism aggregates first across times and only then across persons. It first calculates the aggregate good in each life considered as a self-contained unit and then combines these measures across lives. The alternatives—aggregating first across persons at a time or simultaneously across persons and times—would abandon perfectionism's traditional focus on the life as a morally significant unit (2.2.2, 5.2.2) and in my view make for a less attractive morality.[2]

These assumptions allow us to define some terminology. Any neutral consequentialism tells us to promote a certain quantity, which we can call *global value*. If perfectionism aggregates first across times, it takes global value to be some function of a quantity calculated within complete lives, or of their *lifetime value*. Lifetime value is a function of the essence-development or *perfection* humans achieve at particular times. Chapters 7 through 10 will discuss perfection at a particular time; our present concern is how this perfection is aggregated across times into lifetime value and then across persons into global value.

6.1 Summing and Averaging

6.1.1

The most familiar aggregative principles are *summing* and *averaging*. A summing perfectionism equates each person's lifetime value with the sum of her perfections at all the times in her life, and global value with a sum of lifetime values. Its overriding goal is therefore the greatest sum of perfection by all humans at all times. Of course, much of this global sum is beyond any particular agent's power to affect, but when an agent acts rightly, it is always by increasing the global sum.

An averaging perfectionism equates the aggregate good in a collection with the mean good per member. If it aggregates first across times, its goal for each life is the greatest average perfection per day or year lived and its overriding goal the greatest average lifetime value in history. Because it averages twice, its global goal is an average of averages: the highest mean possible of the mean perfection per year in all lives lived. We are to maximize the average good per life, where this itself is an average good per year.

Summing and averaging support the same moral judgements when the lengths of lives are held constant or we are not adding to or subtracting from a population. But they can diverge otherwise. Then summing favours adding to a collection until its sum of goods is the greatest possible, whereas averaging prefers the often smaller collection with the highest average good per unit.

Although both principles have attractions, each has also been thought open to objections. Against summing there is raised the *repugnant-conclusion* objection, which in its single-life form runs as follows:[3] Try to imagine an ideal finite human life, one that lasts a tremendously long time and has a tremendously high level of goodness throughout. If summing is correct, there is another possible life that is better, even though its level of goodness is never more than barely positive. If this life is sufficiently long—if it contains enough extra years—its sum of goodness will be greater than in the supposedly ideal life. Many philosophers find this claim counter-intuitive, as they do a parallel claim across lives. Try to imagine an ideal course of human history, with a tremendous number of lives of tremendous lifetime value. Summing prefers a different history with many more lives all barely worth living.

This objection does not touch averaging, which never sacrifices quality per time or life for quantity. But averaging has been thought open to a different, *mere-addition* objection.[4] Imagine that a person has led an extremely good life, with an extremely high average goodness per year, but in the future her life will be slightly less good. Her coming years, although still high in whatever makes for goodness, will be below the average for her life to date. Summing thinks this person should continue living, but averaging does not. It says that, because additional years would lower her average per year, it would be better if they were not lived. Many philosophers find this claim hard to accept. How, they ask, can adding something in itself good make a life worse? They react similarly to a parallel claim across lives. Averaging says that if future generations will have less good lives than those lived to date, we should cease procreating. It makes an even stronger claim about the past: that it would have been better if all but the best lives had never been lived, and that this could have been better even if those best lives were thereby made less good.

These two objections can be stated independently of the content of a consequentialist theory, but their persuasiveness is not independent of that content. In particular, it is not the same given utilitarian and perfectionist contents.

Utilitarians who consider the repugnant-conclusion objection have mixed reactions. Some find the objection decisive against summing, but others are prepared to face it down and affirm that, in the first version of the objection, the longer life is indeed better. Perhaps both reactions are understandable. Facing down the mere-addition objection would not be understandable, however, because with utilitarian values this objection does seem unanswerable. When the good is satisfaction, it is not credible that more of it—more pleasure or desire-fulfilment—can make a life or history worse.

Perfectionist values change the situation. Now the repugnant-conclusion objection is, if possible, even more telling. However hard it is to accept a sacrifice of quality for quantity with pleasure or desire-fulfilment, it is even harder with goods of excellence, such as knowledge and achievement. (Think again of Achilles. Surely the loss of his greatest feats could not be made good by any number of successful shoelace-tyings.) At the same time, the mere-addition objection has less force, because with perfectionist values we often do balk at mere additions. Consider our attitude to careers. Many of us think that Muhammad Ali's boxing career, for example, would have been better if it had not contained Ali's last fights against Larry Holmes and Trevor Berbick. Our view is not that Ali's performances against Holmes and Berbick were by some impersonal standard bad. We know that, for many another boxer, to do as well as Ali did against these opponents would have marked the pinnacle of his career. It is rather that Ali's last fights were so far below the level of his prime that it was bad for *his* career to contain them. By falling below his own prior achievements, they made his career worse.

Our view is not that Ali's last fights somehow deprived his earlier ones of value; that did not and could not happen. For many of us, however, the value of a career does not depend just on the (sum of the) values of its parts. It depends also on the career's shape, and the same great feats followed by a long decline can make for a worse career than if they led to an early retirement.

These points suggest that averaging may be more plausible in perfectionism than in other consequentialist theories, and I think it is. We can understand averaging perfectionism as extending to human lives, and then to all human history, an attitude many of us take to careers. Often this extension is appealing in itself.

Imagine that medical technology one day allows people to live much longer than we do now, say, for 150 years. If it does not give them more years of high-quality activity—if it merely extends their old age without adding to their prime—many of us will wonder whether that technology is worth using. In part, we can imagine better uses for society's resources, but I think we will also question the value of the lives the technology supports. Even if they contain nothing intrinsically evil, their period of decline seems out of proportion to their time of peak achievement. Like a play, a life can have a denouement, but it should not take up two thirds of its running time.[5]

The extension across lives also has some attractions. Human history is less like a career than is a single life because it is less tightly unified. But we can at least understand the view that, if the human species has always lived at a high level of excellence, it would be bad for it to continue at a lower level. We can even understand averaging perfectionism's claims about the past. A boxer's career might have been better if it had contained only his best fights and no mediocre ones; the resulting consistency would have been impressive. In the same way, human history might have been better if it had contained only consistently excellent lives.

6.1.2

Not only is averaging more attractive given perfectionist values, but also something like it is endorsed by some perfectionists. Xenophon reports that Socrates accepted death at his trial because he saw his mental powers were failing and did not want to live through an intellectual decline.[6] Commentators often dismiss this story, but it fits several remarks about death in the Platonic dialogues. Consider the *Republic*'s claim that death is not an evil, especially for a good person.[7] Without great stress on immortality, this claim is hard to sustain given most aggregative views, but it follows naturally from averaging. If a good person has already lived as well as he can each year, he cannot increase his average in the future and his death is no deprivation. But a bad person can always improve his life, and by preventing this improvement his death does him harm.

A clearer approach to averaging is in Nietzsche:

> Many die too late and a few die too early. The doctrine still sounds strange: "Die at the right time!"
>
> Die at the right time—thus teaches Zarathustra. Of course, how could those who never live at the right time die at the right time? Would that they had never been born! Thus I counsel the superfluous. . . .
>
> One must cease letting oneself be eaten when one tastes best: that is known to those who want to be loved long. . . .
>
> All-too-many live, and all-too-long they hang on their branches. Would that a storm came to shake all this worm-eaten rot from the tree.[8]

This passage suggests something like averaging across times, when it says that people who live past the point where they "taste best" live "all-too-long," and it may express a similar view across persons when it says that, because of "superfluous" people, "all-too-many live."[9] These claims might be consistent with summing if Nietzsche recognized intrinsic evils, states that not only lack positive value but also have negative value. But, as I will argue (8.1.2), Nietzsche cannot recognize intrinsic evils. And he seems to care most that the most perfect individuals die at the right time, and they are in least danger of falling below a zero. On the simplest reading, his talk of timely death suggests something close to an averaging view.

6.1.3

To be plausible, averaging perfectionism should not forbid too many mere additions, but here it is helped by some empirical facts about human development. More so than satisfaction, perfection is a good that is present to different degrees in different periods of people's lives. For their first twenty or so years they achieve only limited excellence, but acquire the tools they need to pursue it later. Then come forty or fifty years of high achievement, followed by a final decline. Given the long initial period of low perfection, averaging will not want a person's life to end when she first declines, but only later, when the perfection she can achieve in a year drops below the average for her life as a whole. A similar point applies to the species. Because its earliest members exercised only limited rationality, the average for its history will drop only some time after the species's peak is passed.

To forbid even fewer additions, we could move to a *diminishing marginal value* principle intermediate between summing and averaging. This principle agrees with summing that, of any two lives at the same average perfection per year, the longer is always the better. But it says the value of an additional year in a life gets smaller the more years the life contains and diminishes asymptotically towards zero. The principle's claims about lives are represented in Figure 6.1, which shows how a person's lifetime value is a function of the years she lives, given a fixed average perfection

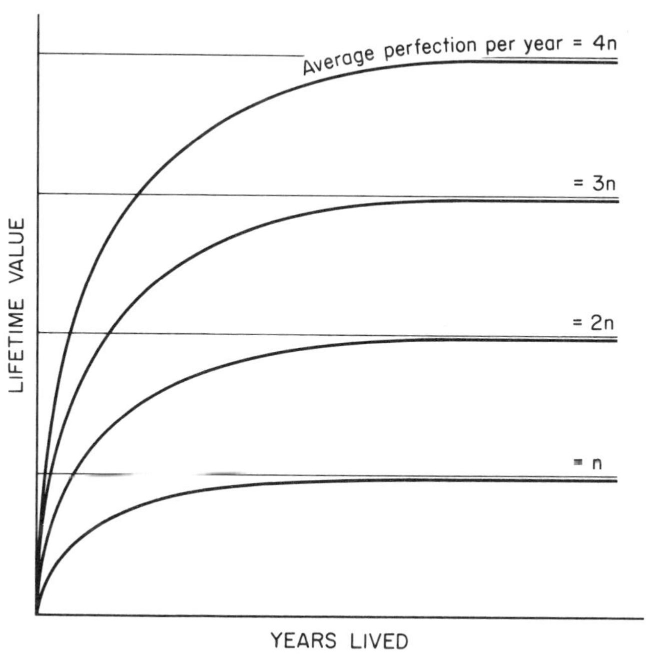

Figure 6.1

per year. By placing an upper bound on value at any fixed average, the principle joins averaging in avoiding the repugnant conclusion. Because it gives some weight to years lived, however, it forbids fewer mere additions than does averaging. If it tells a person to end her life, it will usually be after averaging has done the same.

These claims about lives seem close to Aristotle's. Whereas Plato denies that death is an evil, Aristotle calls it "the most terrible of all things; for it is the end,"[10] and worse for good people than for bad because it deprives them of more.[11] More generally, he holds that excellence requires "a complete life,"[12] by which he means a life of a certain duration. But Aristotle does not yearn for lives that last hundreds of years. His view seems to be that it is important to live for some years, those of a normal human lifespan, but no tragedy not to live beyond this. This view is close to what the diminishing marginal value principle implies.

Applied across persons, this principle says the value of an additional human life gets less the more lives there are. It again avoids the repugnant conclusion while moderating averaging's claims about mere additions. The existence of less good humans in the past, for example, is bad because it lowered the average lifetime value but good because it increased the number of lives, and sometimes the second consideration outweighs the first, especially when the existence of the less good helped increase the perfection of the best. Across as within lives, diminishing marginal value retains the merits of averaging while avoiding some of its defects: For this reason, it is probably the most attractive principle of this general type.

6.1.4

Is the best perfectionism then an averaging-style perfectionism? Averaging and especially the diminishing marginal value principle make some attractive claims about aggregate values but are problematic in a pure perfectionism.

Averaging perfectionism can tell people to end their lives if their level of perfection will be lower in the future than it was in the past. If they refuse, it can tell others to end their lives for them. But that killing people without their consent is wrong, even to make their own lives better, is a firm and widely shared conviction. Nietzsche may be willing to flout it—witness his call for a storm to come and "shake all this worm-eaten rot from the tree"—but we can hardly do the same if we wish to develop a perfectionism in harmony with common moral judgements. For us its implications about killing are a serious objection to any averaging-style view.[13]

A defender of the view may say these implications will rarely be actual. Given the empirical facts about human development, averaging and especially diminishing marginal value will not often tell actual people to end their lives. Their main practical consequences will follow from their application across lives, where they tell us, plausibly, not to increase the human population if doing so will lower the average lifetime value. What is more, if it is an aspect of Aristotelian perfection to make autonomous choices about one's life (11.1), then there is a further objection to forced killing: Any good it does must be weighed against its specific harm to autonomy. Finally, there are arguments about the possibility of miscalculation—of deciding someone ought to die when they should not—and about the effects of publicly allowing killing. Even if the acts this policy licenses can in principle be justified, the fear it causes may be so great that its net effect on human excellence is negative.

How we assess this reply depends on a more general issue. Objections about killing are not unique to averaging perfectionism but confront consequentialist theories in general, which can permit killing when it does enough to promote, for example, the good of others. If we think arguments like those about miscalculation and fear are sufficient to defend consequentialism in general against objections about killing, we may take the same line with averaging perfectionism. We may retain the theory's attractive claims about intrinsic value and not worry about its implications for killing. If we do not find these arguments persuasive, however, we have two options. One is to retain averaging but constrain the duty to promote the values it defines by independent principles about utility or rights. This option—to include perfectionism in a pluralist morality (1.4)—has always been a possibility. It is especially germane here, where some attractive principles for evaluating outcomes have disturbing implications for action. The other option is to retain pure perfectionism but seek a different aggregative principle. This principle is unlikely to be summing, which is counter-intuitive in repugnant-conclusion cases. But it may be some other, as yet undiscussed, principle.

6.2 Maximax

When applied across persons, summing and averaging are both distributively neutral, giving equal weight to equal gains in the lifetime value of all people. A third principle rejects this distributive neutrality, in an anti-egalitarian direction. Because it is the opposite of Rawls's maximin principle, we can call it *maximax*.

6.2.1

According to maximax, each agent's overriding goal should be not a sum or average of lifetime value, but the greatest lifetime value of the single most perfect individual or, if perfections are not fully comparable, of the few most perfect individuals. There is a single goal for all agents to aim at, but not all agents figure in it. Global value is determined entirely by the good of the few best individuals.

Like averaging within lives, maximax across them finds its clearest expression in Nietzsche:

> "Humanity shall perpetually work at producing individual great men—this and no other is its task." How much one would like to apply a lesson to society and its goals, a lesson that can be learned from the observation of any species in the animal or plant world, that it is only concerned with the individual higher specimen, the more unusual, more powerful, more complicated and more fruitful specimen. . . .
>
> Oh Philistine, as if it would make sense to let numbers be decisive when it is a question of value and meaning! For after all, the question is this: How does your individual life receive the highest value and deepest significance? How is it least

> wasted? To be sure, only by living for the benefit of the rarest and most valuable specimens.[14]

This exclusive concern for the "most valuable" is central to Nietzsche's ethics. It is a "basic error," he says, "to place the goal in the herd and not in single individuals! The herd is a means, no more!"[15] The concern also explains his complex view of egoism. For Nietzsche, egoism is neither always good nor always bad: It has "as much value as the physiological value of him who possesses it." If someone's life promises great perfection, he has an "extraordinary right to egoism"; if not, "the first demand of fairness is for him to take as little space, force, and sunshine as possible away from the well-constituted."[16]

Again like averaging, maximax is more plausible for perfectionist than for utilitarian values. No one proposes maximizing the satisfaction of the single most satisfied individual, but a maximax approach to global value extends judgements we make in other domains of excellence. When we evaluate a civilization or cultural epoch, such as ancient Greece or the Baroque, we often care most about the achievements of its best writers and sculptors and less about its middling or minor figures. In the same way, our judgements about artists' careers commonly focus on their finest works. Imagine that one novelist has written more and on average better novels than a second, but the second has written one novel that is finer than any produced by the first. We can understand the view that the last fact makes the second novelist's career on balance better. By extension, we can understand the view that thinks similarly about societies, making their aggregate excellence depend on the excellence of their most outstanding members.

Although Nietzsche often uses maximax across lives, I know of only one passage where he considers it within them.[17] As we have seen (6.1), his preferred principle across times is something like averaging. But a thoroughgoing maximax theory applies maximax twice, equating each person's lifetime value with the perfection she achieves at her most perfect time or times. Just as society should seek the excellence of its best individuals, these individuals should concentrate on their best moments, ignoring what precedes or comes after. A thoroughly maximax perfectionism gives agents a highly particularized goal: the highest momentary achievement of perfection by any human at any time.

When it is applied doubly, maximax addresses all the cases that distinguish summing and averaging. It avoids the repugnant conclusions, for it refuses to accept any decline from a peak of excellence; but it is uninterested in mere additions. If a life, for example, has already had its best moments, nothing now or in the future can alter its aggregate worth, and it does not matter whether it is added to or not.[18] The same holds for the past: If the existence of less good humans had no effect on the best humans, it was neither good nor bad. If it improved the lives of the best, however, it was good even if it lowered the average value per life.

Some interpretations of maximax try to mute its radical claims. Walter Kaufmann argues that Nietzsche's interest in the best individuals derives from his belief that most lives have zero value, and "no addition of such zeroes can ever lead to any value."[19] But this interpretation cannot be correct. If the will to power is "the innermost essence of being,"[20] it must be realized to some degree in all human lives, and if power is the criterion of excellence, they must all have some lifetime value (8.1.2). Nietzsche sometimes recognizes this implication and says a society built on his principles will require sacrifices from the less perfect. It is "characteristic of a good and healthy aristocracy," he says, that it "accepts with a good conscience the sacrifice of untold human beings who, for its sake, must be reduced and lowered to incomplete human beings, to slaves, to instruments."[21]

Variant maximax-style principles are also possible. A *lexical maximax* principle says that agents should aim first to maximize the lifetime value of the best individuals and then, if nothing more can be done for them, of the next best individuals, and so on.[22] It treats each level of lifetime value as having infinite value relative to any lower level. It also points to a weaker principle intermediate between lexical maximax and summing, one giving just finitely greater weight to higher lifetime values. This *finite priority* principle says the value of a unit increase in someone's lifetime value is greater the more lifetime value he has but is never infinitely greater. There is some reason to concentrate on the best individuals, but not an absolute reason.

Less radical than simple or lexical maximax, this finite priority principle may better match our judgements about epochs and careers, which do give some weight to lesser achievements. It also supports a view that is understandable given perfectionist values: that the duty to develop one's talents, which applies to all humans, is stronger for those with greater talents (5.1.1). But the principle has the defect of again entailing the repugnant conclusions. Simple and lexical maximax never trade quality for quantity, but if less perfect lives or times have even a little value compared to better ones, enough of them can make good any loss of seemingly ideal lives or moments.

6.2.2

Although it extends some plausible judgements, maximax is problematic in pure perfectionism. Technically it is agent-neutral because it assigns the same moral goal to all persons. But it does not have the consequences we normally associate with agent-neutrality. It tells the least perfect to care only about the excellence of a few other people, and those few to care only about themselves. So it hardly captures our judgements about the morally proper treatment of others (5.2.3). Equally importantly, maximax has disturbing distributive implications. Many of us have a broadly egalitarian view about economic distribution: We favour roughly equal distributions of material resources, ones that give everyone a chance at a valuable life. Maximax, however, rejects this view. Caring only about the most perfect, it wants as much wealth, power, and opportunity for them as they can use. The majority get only what is left over, or what will help them serve the elite. Both these consequences are mitigated by the finite priority principle, which requires the best to care to some extent about the excellence of the less perfect and is less radically anti-egalitarian. But it places fewer restraints on the best than on the less good and can still recommend very unequal distributions.

So we face a choice. On one side is an aggregative principle with some affinity for perfectionist values; on the other some broadly egalitarian intuitions. In a pluralist theory we might retain both, weighing maximax perfectionism against some non-perfectionist egalitarian principle.[23] In pure perfectionism, however, we must choose.

Some may say this choice is illusory: Perfectionism has always been anti-egalitarian, they may argue, and talk of its being otherwise ignores its essential character.

This argument is too hasty. Although perfectionism has often been anti-egalitarian, the tradition contains important defenders of distributive equality, for example, Marx and Green. And whatever a theory's final claims about distribution, the way it derives them depends crucially on its structure. Maximax offers the shortest route to elitism because it supports unequal shares on almost any assumptions about the world. But a distributively neutral principle does the same only given certain contentious empirical theses. Some perfectionists accept these theses, but others such as Marx and Green do not, and their broadly egalitarian perfectionism is a possible moral position. The issues here are important. If pure perfectionism cannot favour distributive equality, there will be strong pressure to supplement it with a non-perfectionist egalitarian principle. Issues about distribution are discussed further in chapters 12 and 13, but here we can say this: If we are drawn to a broad egalitarianism, we will find plausible only those versions of pure perfectionism that have a chance of endorsing it. We therefore cannot accept maximax, which has elitism built into its formal structure, but must prefer a distributively neutral principle.

Nor does the case against maximax turn only on independent intuitions. Like averaging, maximax extends judgements we make about, for example, careers. But where averaging retains much of its intuitive appeal when applied to properly moral subjects, maximax does not. In evaluating graduating classes or history departments, we may look especially at their best members. But we hardly do the same when framing social policy for a whole nation. There everyone's excellence matters, and matters significantly. In the nineteenth century, British perfectionists were leaders in the movement for mass education, campaigning for state-funded schools, founding provincial universities, and participating in workingmen's education.[24] They were not content with the accomplishments of a few students at Oxford and Cambridge, but wanted knowledge spread as widely as possible. Many of us share their ideals today. We believe a major task facing the world is to further the spread of literacy, and with it the capacity to understand important facts about our universe and our species's history. We want all humans to live informed, intelligent lives, not just some elite. This may not count against every view with a maximax flavour, such as the finite priority view. But it does show that our most serious judgements reject any exclusive concern with the best.

Why the divergence between our judgements about careers or graduating classes and societies? In the first case, we judge individual achievements in abstraction from the lives in which they figure; in the second, we focus centrally on lives. This focus pulls us away from maximax and towards distributive neutrality. When what is at stake is the goodness of complete human lives, we more naturally count equal gains in all lives equally.

The case against maximax should not be exaggerated: As I have argued, it is an understandable principle for aggregating perfections and has some attraction. This is significant because the same principle is not understandable for aggregating satisfaction. And it helps to explain the special quality of our reaction to Nietzsche. His brazen anti-egalitarianism fills many of us with horror, but this is the kind of horror that involves as an essential element some attraction to its object. We are not horrified at the idea of maximizing the satisfaction of the few most satisfied individuals, for the idea is a non-starter. But Nietzsche's view strikes us as dangerous, because somehow dangerously attractive.

This point has a further implication. It may be suggested that, if we favour distributive equality, we should adopt a positively egalitarian aggregative principle, for example, one that gives finitely more weight to gains by the less perfect. This, it may be said, is Marx's approach: to care more about increasing the excellence of the least excellent than augmenting further the accomplishments of the best.

I do not believe this move is plausible in perfectionism. (Nor is it the best explanation of Marx's egalitarianism; see 13.1.) Formally egalitarian principles are attractive for utilitarian values and are often combined with them, but they do not fit perfectionist values. It is not plausible to prefer a small gain in excellence by the least excellent to a large gain by the best, for example, a small gain in musical achievement by a beginning music student to a large gain by Mozart. We can weigh unit gains to all people equally, but it violates our intuitions about excellence to prefer gains lower down a scale of development. Just as our notions of excellence and perfection exclude satisficing or aiming at less than the best (5.1.1), so they exclude caring less for the best. In utilitarianism the intuitively understandable principles run from distributive neutrality to extreme egalitarianism, and anti-egalitarianism is a non-starter. In perfectionism the situation is reversed. There the understandable principles run from neutrality to anti-egalitarianism, and egalitarianism is not plausible. It does not follow that perfectionism must be formally anti-egalitarian: It can and should be distributively neutral. But there is a constraint on any attempt to reconcile perfectionism with distributive equality. The attempt cannot use formally egalitarian principles but must combine neutral aggregations with empirical claims about what most promotes the equally weighted perfection of all (see further chapters 12–13).

6.3 Single-Peak Perfection

6.3.1

A fourth principle is proposed by Thomas Nagel. He argues that with perfectionist values we care only about the "level" and not the "spread" of human achievements.[25] When we consider feats like climbing Mt. Everest or landing on the moon, we care especially that they be performed once but much less if they are repeated by others. And if there are obscure truths, for example, about Portuguese politics in the fifteenth century, we want someone to know them but do not mind terribly if the knowledge is not widely shared. As Nagel says,

> It is important to achieve fundamental advances, for example, in mathematics or astronomy, even if very few people come to understand them, and they have no

> practical effects. The mere existence of such understanding, somewhere in the species, is regarded by many as worth substantial sacrifices.[26]

Let us call this the *single-peak perfection* principle. In its pure form it finds value only in the first climb of a mountain or first knowing of a truth, and no value in repetitions. A weaker principle finds just finitely more value in first performances, with the value of repetitions perhaps diminishing to zero.

Like maximax, the single-peak principle disregards numbers, but it does so in a different way. Imagine that a person absorbs much of the knowledge acquired by others and duplicates many of their greatest feats in different areas. A maximax theory may say his lifetime perfection is greater than anyone else's and makes all others' doings redundant; on the single-peak view, his achievements, if they contain nothing new, are valueless. And even if the achievements are new, they still leave work for others: If there are feats he has left undone, there is value in others' performing them. What maximax refuses to add up is the lifetime values in less than the best lives; single-peak, by contrast, refuses to add the particular values in different instances of the same perfectionist feat.

Beyond individual feats, single-peak can be applied to whole activities or forms of life. We can care especially that our country have at least one high-quality newspaper or support some opera, and we can think it more important to preserve an activity like opera than to extend or improve others that are less threatened. Our view need not be that opera is better than these other activities. Even if it is just one perfection among many, we can care more about having some of it than about having more of what is already flourishing.

6.3.2

Like other aggregative principles, single-peak can be applied across times as well as persons. Then it says that just as a second person's achieving a good now adds no value now, so the good's being achieved later will add no value later. This application may be plausible for feats such as climbing Mt. Everest, although I suspect many would find the climb newly important if no one had done it for years. But it does not fit our view about knowledge. Even when we seem to use single-peak at each time and think one knower per time is sufficient, we care that there be such a knower at every time. We have a picture of human knowledge as something that persists through time and is diminished when old information is lost. Without this picture we could not value knowledge of the past. If fifteenth-century Portuguese politics were known in the fifteenth century, we can only care that someone know them now if our principle across times is other than single-peak. The same goes for activities like opera: We can only care that there be some now if we do not think a season a century ago is enough.

To be plausible, then, a single-peak theory must often apply its principle only across persons, but then it cannot aggregate across times before persons. Instead of first calculating the value in each life considered as a unit, it can evaluate an individual state only after looking first across persons at the other states simultaneous with it.

In my view, this reflects a tension between single-peak and the foundations of narrow perfectionism. If we see the good as grounded in human nature, we will naturally think of it as achieved within whole human lives. We will see value as coming in a unit that single-peak ignores. Nonetheless, single-peak remains important to discuss. Like averaging and maximax, it is more plausible for perfectionist than for utilitarian values. Although no one holds that once one person has experienced the pleasure of, say, eating chocolate it is irrelevant whether others do, the parallel view about perfections has, as Nagel shows, some attraction. And this may pose a challenge to narrow perfectionism. If the most attractive principle for aggregating perfections broadly conceived is in tension with the narrow ideal of human nature, may this not tell against the ideal?

6.3.3

Nevertheless, single-peak is not the most attractive principle. Our tendency to endorse its judgements can often be explained on other grounds, and when it cannot, the tendency is not morally serious.

With feats such as climbing Mt. Everest, our interest in first performances often reflects their greater difficulty. Someone who climbs a mountain first cannot know how others have done so or even if the feat is possible. This makes his climb a greater achievement and, on any aggregative view, gives it more value. (If, unbeknownst to each other, two climbers ascend a peak at the same time, is either's achievement diminished?) As for preserving knowledge and opera, these have obvious instrumental justifications. If someone knows Portuguese politics today, it is possible that others will know it in the future. If we support at least one opera company, a tradition will continue that can enrich future generations. Instrumental considerations go a long way towards justifying preservation, and when they are controlled for, single-peak judgements are harder to sustain. Imagine that we have a choice between preserving opera and making improvements in the elementary education of all children. Imagine further that the net effect through time on the sum of human excellence of the educational improvements will be greater. I think we will judge it right to prefer the educational improvements. We will recognize the loss of opera as a loss, but one that in the circumstances is morally justified.

These considerations may not account for all of our interest in unique performances, but what remains is of questionable moral significance. It is a *Guinness Book of World Records* attitude that is not as serious as our commitment to ensuring literacy, knowledge, and rationality for all (6.2.2). This is partly because single-peak aggregates achievements in abstraction from the lives in which they appear. When we focus on lives, as narrow perfectionism forces us to, our most serious judgements give equal weight to unit gains in all human lives.

6.3.4

A view connected to single-peak has been defended by Derek Parfit.[27] Concerned to avoid the repugnant conclusion, Parfit proposes giving infinite relative weight to certain individual states higher up a continuous scale of perfection. Listening to

Mozart's music, for example, has the same virtues to a higher degree than listening to Muzak. Nonetheless, Parfit argues, the experience of Mozart should count infinitely more, so its loss could not be made good by any increase in the quantity or quality of Muzak.

The simplest version of Parfit's view gives *any* experience of Mozart infinite weight relative to Muzak, but this is extremely anti-egalitarian. It implies that even if there are already many people experiencing Mozart, it is more important to give someone who has heard a hundred first-rate performances a ticket to yet another performance than to make large improvements in the lesser perfections of millions of badly off people. Like maximax, although for a less plausible reason, the view can require us to concentrate benefits on those whose lives have already been best.

The more credible versions of Parfit's view combine it with something like a single-peak principle. One such version says that what has infinite relative value is just the existence of some Mozart in the world, either at each time (if the relevant principle is applied only across persons) or at some time (if it is applied doubly); once one person experiences Mozart, the addition of further listeners has just high (finite) weight. This version differs from single-peak only in giving the existence of extra listeners less (rather than no) weight and in making a strong claim about a topic single-peak does not address: the relative value of having some of a higher perfection versus some (or a lot) of a lower perfection. A different version of Parfit's view applies an analogue of single-peak within lives. It says the value of one person's listening to Mozart is not affected by facts about whether others are listening or have listened to Mozart. What has infinite relative value in her life, however, is just that she experience some Mozart, at some time. Once she has listened once to Mozart, the addition of further Mozart in her life has only (high) finite weight.

These latter versions of Parfit's view are less anti-egalitarian than the simple version, but they still have disturbing implications about distribution. They imply that if only one person can appreciate Mozart, it is better to spend vast quantities of resources to enable her to do so than to share the resources equally, which is hard to accept if we have broadly egalitarian intuitions. And this failing reflects the general counter-intuitiveness of all single-peak–style views. Distribution aside, do we really believe that one experience of Mozart, either in a life or in the world, is worth unlimited sacrifices of lesser perfections such as literacy?

Finally, as Parfit himself notes, the view he proposes is structurally odd. It can be understandable to give one good infinite weight relative to another very different good, for example, autonomy versus pleasure. I believe it is also understandable to give infinite weight to something as unified as the best human life. But it is odd to do so for a value that merely comes higher on a continuous scale: If Mozart differs from Muzak only by degree, how can its value be different in kind? Parfit's view has the merit of avoiding the repugnant conclusion but is hard to square with a scale of continuously increasing Aristotelian perfection.

6.4 Conclusion

We have examined several aggregative principles and found none entirely satisfactory. Summing weighs quantity too heavily against quality and entails conclusions that are especially repugnant given perfectionist values. Averaging and especially the diminishing marginal value principle avoid these implications and make some attractive judgements about states of affairs. But they are problematic when applied directly to actions, for they can require killing people against their will. The other principles, maximax and single-peak, have some attraction given perfectionist values, but they do not fit our most serious perfectionist thinking and, especially maximax, violate intuitions favouring equal economic distribution.

We could try using one principle across times and another across persons, for example, using diminishing marginal value within lives and maximax across them. But this would not avoid the difficulties because no principle has proved immune to objection in any of its applications. We could also move to a pluralist theory in which perfectionist values, aggregated in some way, are weighed against principles about rights or equality. But this would be to abandon our project of developing a pure perfectionism. Within that project, there seems no single unobjectionable aggregative principle.

Although it is disappointing, this conclusion should not be surprising: Questions about aggregation are difficult for any consequentialist theory.[28] Perfectionist values change the aggregative options, making some principles more attractive than they are given utilitarian values, and others less attractive. However, they do not point to a single, simply acceptable principle.

Still, our conclusion is not completely empty. Even if we have not selected a single determinate principle, we know many of the properties an acceptable principle must have. Unlike single-peak, this principle must calculate values first within human lives. Unlike maximax, it must be distributively neutral, giving equal weight to equal gains in all lives. So in its general character it must be similar to summing, averaging, and the diminishing marginal value principle. And we can expect this principle, if it exists, to agree with summing and averaging in the cases where they agree, when we are not affecting the length or number of human lives. If we restrict ourselves to these cases, we can say the most attractive perfectionism equates aggregate value with something like a sum or average of perfection. This conclusion cannot be extended beyond these cases, but seems adequate for them.

7

The Well-Rounded Life

To aggregate perfection, we must be able to measure it—to say how far each human has developed her nature at each time. We must not, however, expect either too much of perfectionist measurement, or too little.

We expect too much if we think perfection can or must be measured on a precise cardinal scale, one that assigns exact numbers to all human states. Such precision is not possible for perfectionist values. We can often say that one activity is more perfect than another and by roughly how much, but we cannot go beyond this.

We expect too little if we assume that, without precise cardinality, we can say nothing systematic about measures of perfection. Even without exact numbers, we can ask what specific states contribute to the different excellences, and how. We can compare different views about how the elements of human nature are realized. And in evaluating these views, it is often helpful to ask what mathematical principles would be best for measuring perfection *if* precise cardinality were possible. This last claim is well illustrated by our topic in this chapter.

The measurement of perfection involves two tasks. First, we explain what counts for and against a person's achievement of each perfection considered on its own. Second, we explain how her combined perfection at a time arises out of her achievement of the different perfections taken together. This second task is necessary because a person can face choices among perfections. On a particular day, she may be able to read a history book or work in a political campaign. Here she can increase her theoretical or practical perfection but not both, or not both equally. More generally, she must choose between a life of predominantly theoretical achievement—an intellectual life, if you like—a life devoted to action, and a life that tries to balance the two. After characterizing the individual excellences, Aristotelian perfectionism must help a person make these choices. It must explain how her combined perfection at a time arises out of her achievement of the different perfections, or how she can compare perfections when they conflict.

This chapter discusses the second element of perfectionist measurement, using a simplifying assumption. It assumes that we have a precise cardinal measure of each perfection on its own and asks how these measures are combined. As before, the assumption is unrealistic, but we will soon learn why it is helpful.

7.1 Lexical and Constant Comparisons

How should perfections be compared? So far as I can see, the general perfectionist idea imposes no restrictions here, nor does the Aristotelian theory of human nature. Together they yield a list of perfections, but say nothing about how they are weighed. The best perfectionism can use whatever comparative principle seems most attractive.

7.1.1

For many perfectionists this principle is very simple. They say there is one perfection, often theoretical perfection, that we should always seek ahead of others. If there are different duties to pursue different goods, they give one duty *lexical* priority. Aristotle takes this line when he ranks contemplation ahead of any practical perfection. He wants us to "strain every nerve" to develop our theoretical faculty, implying that we should prefer any intellectual activity to the exercise of practical reason.[1] Aquinas has a similar view about the relation between rationality in general and bodily development. Unlike Aristotle, Aquinas thinks physical perfection has some value,[2] but this value is "ordained to" spiritual goods and must always yield before them.[3] Both writers make one perfection infinitely more valuable than the rest; even a tiny amount of it outweighs the greatest achievement of them. Only when we can no longer promote the higher excellence should we seek a lesser good.[4]

Common though it is, this lexical approach is too extreme to be plausible. Each Aristotelian perfection has some attraction, and none deserves to be placed so far above or below the others. Physical perfection seems to me less important than the two rational excellences (4.1.2). I have no special argument for this, but the development of our bodily powers seems to have less value than the exercise of rationality. So in my view it should count for less. To make any increase in rationality, no matter how trivial, outweigh the greatest bodily achievements, however, is to go too far. A plausible morality must give each Aristotelian good some serious weight.

7.1.2

The simplest such morality employs *constant trade-offs* between perfections. It assigns a fixed, finite weight to each excellence, and this weight is the same for all people in all circumstances: if two units of theory equal three of practice for me, they do so for you and everyone else. Because there is no pre-existing scale on which perfections are measured, this view cannot in the strict sense make one higher. (Halve the units for theory and the above proportion of 2:3 becomes 4:3.) But it can do this informally. If a comparative principle implies that most normally endowed people should spend more time pursuing one good than another, then, informally, it makes that good more valuable. This is just what we should want for rational versus physical excellence.

Constant trade-offs still permit Aristotle's view that theory is better than practice, now read as the non-lexical claim that it deserves more finite weight. But even in this weaker form, the view is implausible. Theoretical and practical perfection both develop rationality, and do so in structurally similar ways (chapters 8–10). Both are products of natural selection, and nothing in their character or origins makes one more desirable than the other. Why should rationality in conduct, in how we change the world, count less than rationality in how we form our beliefs? Why should a structure of ends have less value than a similar structure of judgements? Aristotle's arguments for preferring theory are unimpressive. His central argument—that theoretical excellence realizes a separable and divine element in our nature[5]—is of merely historical interest, as is the claim that it concerns the best objects.[6] Remove God and we have no reason to believe that what theory knows is better than what action affects. His other arguments are no more persuasive. That contemplation offers pure and continuous pleasures is irrelevant in a theory that does not value pleasure; that it is more self-sufficient is dubious (see 8.4), and that it alone is loved for itself is false.[7] Players value skill in games not just as a means to winning—many care little for that—but because in itself it exercises rational capacities. Without better arguments than Aristotle's, perfectionism should give the rational perfections roughly equal weight, so a world of uniformly good lives devotes roughly equal time to each.

7.1.3

However they rank goods, constant comparisons are easy to grasp technically. Given precise cardinal measures of the individual perfections, they yield a precise measure of combined perfection given some precise (constant) weights. But assigning these weights, however simple in principle, would be implausible in practice. Imagine that some precise schedule of trade-offs, say, 2:3, has consequences we find morally appealing. There are many other schedules that differ only slightly in their assigned weights—say, 201:300—and only slightly in their implications for particular cases. Can we really say the first schedule is better than all these others? In choosing comparative principles, we have only our intuitions to go on, either about principles or about particular cases. How could they support such finely grained weightings?

This argument shows that, even with precise cardinal measures of the individual perfections, we would be foolish to try comparing them exactly or arriving at strict numerical values for combined perfection. But it does not follow that we can say nothing at all about comparison. Even if we cannot adopt one specific schedule of trade-offs, there are many we can reject. If we think trade-offs should be constant, we can set aside any that are not; if we have a general sense of appropriate weights, for example, rough equality, we can require that too. Beyond this, we can appeal to our particular judgements. We all have some convictions about cases where perfections compete, and can reject any principles that violate these convictions. We believe that if a person learned some trivial truth in a way that frustrated her major career project or caused her finest friendship to end, her theoretical gain would be outweighed by the practical loss. We also believe that if she won an isolated game of checkers by taking a drug that destroyed all her scientific knowledge, the converse would hold. If we reject all principles inconsistent with judgements like these, we will have left a set of comparative principles that we can say are *acceptable*. And we can use this set to make at least some comparisons of value. We can say that whenever a claim about combined perfection is true on every acceptable principle, it is true *simpliciter,* but that otherwise the claim is neither true nor false. Thus, if state *a* contains more combined perfection than *b* on every acceptable principle, it contains more combined perfection *simpliciter*. If it contains at least twice as much combined perfection as *b* on every acceptable principle, it contains at least twice as much combined perfection. We can make this rough cardinal claim even though there is no precise measure of the value in either *a* or *b*.

This device, which logicians call *supervaluations,*[8] cannot always rank states precisely. If different acceptable principles differ about the relative merits of *a* and *b,* it will not be true that *a* is better than *b,* that *b* is better than *a,* or that *a* and *b* are exactly equal in value. Even so, some cardinal claims are still possible. Perhaps on no acceptable principle is *a* or *b* as much as twenty percent better than the other. If so, it is true *simpliciter* that neither state is as much as twenty percent better than the other: *a* and *b*, although not exactly equal in value, are roughly equal.

If two states are roughly equal in value, the immediate consequence is the same as if they had identical combined perfection: If we can bring about either, it is morally indifferent which we select. But the more remote consequences are not the same. If *a* and *b* are exactly equal in value, then, if state *c* is better than *a,* it must also be better than *b*. If *a* and *b* are just roughly equal, however, then state *c* can be better than *a* yet roughly equal to *b*.

This point is crucial for a realistic perfectionism. If a person can choose between reading thirty pages of philosophy in the next hour and running six miles, many of us will think it a matter of indifference which she selects. But she will do better to read thirty-one pages than thirty, or to run six and a quarter miles than six. Therefore, if her first choice involves a strict equality of combined perfection, she must prefer reading thirty-one pages of philosophy to running six miles, which is absurd. If one choice can be morally indifferent, so surely can another slightly different choice.

Granting this point, why capture it with a formal device like supervaluations? The great merit of supervaluations is that they steer us between the errors of expecting too much from perfectionist measurement and expecting too little. On the one hand, they answer any charge that our lack of precise cardinal measures reflects some incoherence in the idea of combined perfection. Our problem, they reveal, is not that we cannot formulate any plausible comparative principle—nothing is easier—but that we can formulate too many. On the other hand, they allow us to analyze comparison systematically. Even if perfectionism uses many comparative principles, there can be formal properties that any principle must have to count as acceptable. Given supervaluations we can ask what these properties are, and do so just as if we were formulating a single principle. We can ask what the most attractive formal properties are, not to arrive at a single determinate principle, but to place constraints on a set of principles.

What supervaluations allow here, they allow elsewhere in perfectionism. Our discussion of aggregation, for example, presupposed the device. It assumed a precise cardinal measure of combined perfection because, even if precise cardinality is unrealistic, we can ask how perfection is aggregated on all acceptable measures. The same will hold when we discuss the individual perfections. We will consider formal principles for measuring theoretical and practical perfection, not because we can adopt these principles as such, but because they express what is common in our informal judgements. Those judgements are at best roughly cardinal, but we can understand them as arising from supervaluations over a set of precise judgements.[9]

7.2 Balancing

7.2.1

Let us return to the topic of comparison. Our initial suggestion was constant trade-offs between the different perfections, with physical perfection ranked lower and the rational goods roughly equal. This proposal is not positively unattractive, but I think many of us are drawn to something different. We think the best lives contain a certain balance among perfections and do not concentrate too much on any one. To talk only of Leonardo would pitch the ideal too high, for what I intend is possible in all lives. We all can spread our activities widely, aiming at a well-rounded achievement rather than any narrow specialization. Even if our individual accomplishments are not great, their proportion can mirror that of Renaissance lives, and for many of us this proportion is, other things equal, a good. At however modest a level, it gives our life an intrinsically desirable shape.

This *balancing* view, however, excludes constant trade-offs. Imagine that one person has devoted most of her life to politics, while another has worked only at scholarship. An ideal of roundedness implies that the first will improve her life most by acquiring some knowledge, whereas the second should become more active. Even with the same options, the two should choose differently. They should prefer what they have neglected, and what they have neglected is different. At the heart of balancing is the idea that a perfection's relative value depends on the relative amount of it one has achieved in the past. Going beyond equal weights, it says that if one excellence has been achieved more than another, the second is more important. The clearest representation of this idea is on an indifference graph (Figure 7.1).[10] Here each curve links points representing mixes of the two perfections that make for the same combined perfection, with curves further from the origin representing greater

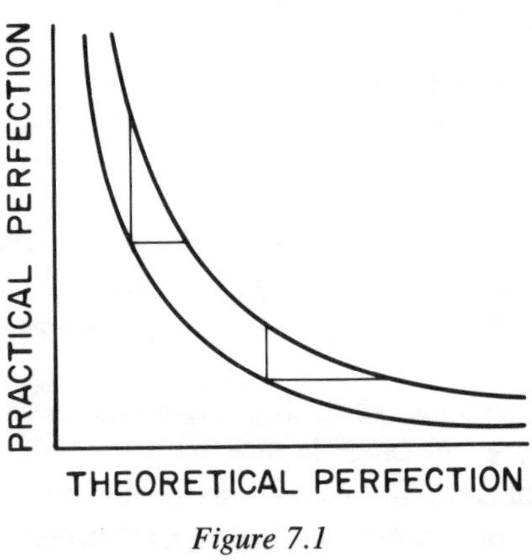

Figure 7.1

values than ones closer in. (For simplicity, imagine the axes calibrated so that a unit of each perfection is what an average person can achieve in a fixed time, say, an hour.) In the upper left of the graph, where the politician has achieved more practice than theory, the same overall improvement results from a large increase in rational action or a small increase in knowledge. So for her, knowledge is more valuable. In the lower right, by contrast, the scholar makes an identical gain by improving theory greatly or practice a little.

Indifference graphs are borrowed from economics, where they represent consumers' preferences among commodities. But there is an important difference between the economists' use and our own. On a natural view, the economists' curves reflect the fact that the absolute utilities of commodities diminish as their absolute levels increase. Thus, the reason we prefer an apple to an orange when we have more oranges than apples is that we care less for either fruit the more of it we have. This cannot be the point in perfectionism. Imagine that one person has ten units each of theoretical and practical perfection, while another has twenty. If the value of balanced lives rested on diminishing absolute values, the second person's life would be less than twice as good as the first's, which we do not want to say. Our aim is to appreciate Leonardo, not to minimize his feats. For this reason, acceptable balancing principles must look only at a person's relative achievements, that is, at ratios between goods and not at their absolute levels.[11]

7.2.2

Given the intuitive appeal of balancing, it is no surprise that it is endorsed frequently in the perfectionist tradition. Several writers say that an ideal life involves a "harmonious" achievement of different goods. Hamilton defines perfection as "the full and harmonious development of all our faculties,"[12] Humboldt says a human's end is "the highest and most harmonious development of his powers to a complete and consistent whole,"[13] and Rashdall describes a life where "many distinguishable elements are harmonized and combined."[14] At times the point of these remarks is causal. Humboldt says one can "increase and diversify the powers with which he works, by harmoniously combining them, instead of looking for a mere variety of objects for their separate exercise."[15] This is sound advice on any view about comparison. If some activities or ways of organizing activities advance two excellences at once, any perfectionism should applaud them. The talk of harmony often goes beyond this causal claim, however, to suggest that some desirable proportion between goods is violated if we concentrate too much on one. This suggestion is explicit in the polemics of Marx and Nietzsche against specialization and the division of labour. Marx wants famously to be able "to do one thing today and another tomorrow, to hunt in the morning, fish in the afternoon, rear cattle in the evening, criticize after dinner, just as I have a mind, without becoming hunter, fisherman, cowherd, or critic";[16] Nietzsche heaps scorn on the "nook-dwellers" and "fragments of humanity" he finds among European intellectuals.[17] Against "modern ideas" that would "banish everybody into a corner and 'specialty,'" he insists that a human's greatness lies in his "range and multiplicity, in his wholeness in manifoldness."[18]

7.2.3

The exact impact of balancing depends on facts about the lives people have available, which we discuss in the next section. But it also depends on the goods' relative weights. Like constant comparison, balancing can make one perfection higher if it tells normally endowed people to spend more time pursuing it. (This will be reflected in curves that tilt towards one axis, rather than lie symmetrically, as in Figure 7.1.) We should take this line with rational and physical perfection. Then our morality will say that, although it is important for a life to contain some bodily goods, there should usually be less time spent on them than on rationality.

Balancing has several indirect consequences, of which one is to reduce the importance of an excited chase after goods. On any view, the best lives strive vigorously after excellence. But with balancing this striving does not guarantee high combined perfection. Many lives of all-out activity fail, through lack of balance, to equal the value of lazier lives whose proportion is right. At the same time, balancing increases the importance of self-knowledge. With constant trade-offs people can compare perfections without knowledge of their past. Because the relative weights of the perfections are always the same, they can choose correctly without knowing how they chose before. But balancing requires awareness of one's history. As an instance of theoretical perfection, this awareness is intrinsically good, and by contributing to a unified life it also enhances practical perfection (9.4). Here we see how, given balancing, self-knowledge is a condition for correct comparative judgement.

Alongside its consequences for the general shape of lives, balancing affects one's choice of specific activities. As Humboldt notes, one way to achieve "harmonious development" is to choose projects that promote several excellences at once. These projects are commended by any perfectionism, but even more so by one with balancing. Scientific research is multiply valuable because it combines the separate goods of theory and practice (8.1.3), and so are many kinds of athletics. Team sports require players to solve strategic problems at the same time as they exercise their bodies. Here the mix of rational with physical goods is what balancing finds attractive and what suits these sports to a well-rounded life.

Given this last point, a life could count as well-rounded, in the sense defined to now, even though it concentrated on just one activity. A life devoted entirely to research could be balanced if its inquiries developed practice as much as theory, and a hockey-playing life could be rounded if the sport's intellectual demands match those on the body. This implication may seem counter-intuitive and false to our original ideal of well-roundedness. It may seem small honour to Leonardo to think his virtues attainable in a life with just one domain of achievement. The implication can be avoided, however, if we extend balancing from different perfections to different realizations of one perfection. Then it will, other things equal, be better to know European history, astrophysics, and a friend's character than to concentrate understanding in a single area, and better to complete diverse projects than to have just one kind of practical achievement. This extension is natural, and I will assume it in what follows. I will assume that we are to seek variety not just among excellences but among aspects of each excellence, so a fully rounded life is developed, knowing, and active in many fields.

7.3 Dilettantism and Concentration

Balancing implies that, other things equal, we should prefer the perfections we have achieved less of in the past. To apply this directive, we must know when other things are equal or, more precisely, what other effects a search for balance will have. Does diversity of action undermine the individual goods, so a well-rounded life will be uniformly mediocre? Then balance, although attractive in theory, is not recommended in practice. Or does diversity enhance the perfections, enriching them from different sources? Then balance is doubly required.

7.3.1

To sharpen these questions, let us return to our indifference graph and add an *achievement line*. The points on this line represent the greatest combinations of perfection a person can achieve given her natural abilities, her resources, the time in her life, and whatever else affects her achievement of excellence. The area under and including the line represents the total lives available to her, and in comparing perfections she in effect chooses a point in this area. If her achievement line is straight and symmetric to the axes (Figure 7.2), it touches the outermost indifference curve at a single central point, which we can call its *ideal-life point*. Given this line, she lives the best life by giving equal time to both goods.

Real-world achievement lines are hardly straight, however, and their shape can affect the impact of balancing. Imagine first an achievement line that is convex to the origin, so its shape mirrors that of the indifference curves (line i in Figure 7.3). Given this line, balance is not such a virtue or specialization such a crime. Even very concentrated lives come close to the outermost indifference curve and with sufficient convexity could touch it. By contrast, a line that is concave to origin (line ii) makes balance more important. Concentrated lives fall further short of the ideal and are more seriously lacking.

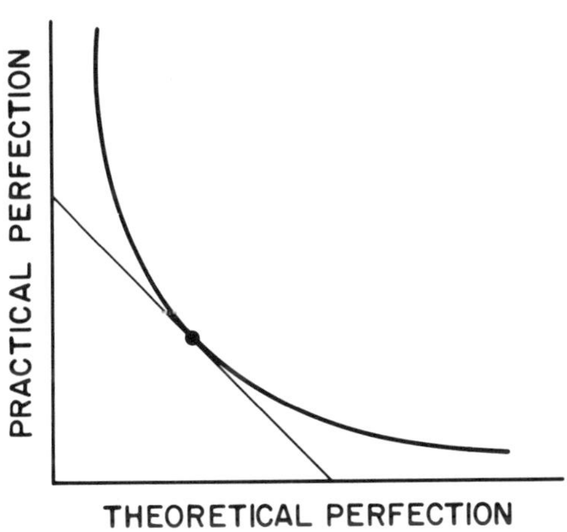

Figure 7.2

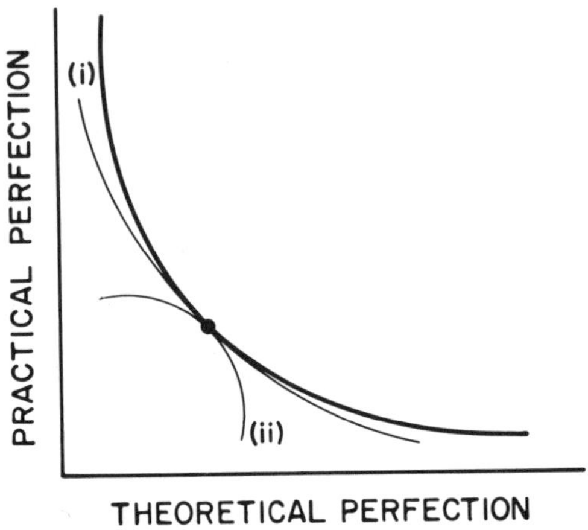

Figure 7.3

Achievement lines belong to individuals, and it would be foolish to think that one shape fits all. Nonetheless, there may be general factors that affect the shape of all lines and determine the general import of balancing.

7.3.2

There seem, unfortunately, to be two opposed factors, one supporting convexity and the other its opposite. In favour of convexity, and against balance, is what we can call *dilettante's disadvantage*. People often find that the more time they invest in an activity, the more they gain from further attention to it. A professional historian can learn more from an hour with a history book than can an athlete, who lacks the background to read it critically. The athlete can achieve more physical perfection in an hour's exercise: He can run further or lift greater weights. In both cases, a person's past devotion pays dividends now. It is often said that those who spread themselves across many activities will not achieve much in any. If this is correct, a well-rounded life will be low in all perfections and, to reflect this, most achievement lines will be convex.

On the other side are the *costs of concentration*. People who devote themselves to one activity can easily become fatigued or bored. They can miss the invigorating effect of variety, as well as the chance to enrich themselves in one area with experiences gained in another. Their very narrowness can produce stultification. What is more, some perfections are subject to a law of diminishing returns that makes it easier to move from low to middling achievements than from there to the highest heights. In ballet, for instance, it takes a certain effort to move from a beginner's level to that of the corps in an amateur production. It takes much more effort to move from there to the technique of a professional production. At the highest levels, dancers practise for hours to improve their movements a little. Given the tremendous costs they incur, might they not do better for themselves if they sought some variety? The worry here is not that the dancers are wrong to practise as much as they do. Given the finer appreciations they permit the many people in their audiences, they are probably right. If the price of those appreciations is such narrow concentration by the dancers, however, may it not leave their own lives impoverished?

These contrary factors are both plausible, and readers may disagree about their relative weights. Some may think the costs of concentration greater, and still value well-roundedness when all is considered. Others may emphasize the disadvantage. But there is a third possibility. We can acknowledge both factors and unite them in one view, if we look more closely at where they are strongest.

7.3.3

Dilettante's disadvantage, to start, seems strongest at low levels of perfection and weaker higher up. In many domains of perfection, large investments of time are needed before any excellence results. In a cognitive field there are basic concepts and principles to grasp, and in a practice there are fundamental skills; without them, a person's activity counts for little. Unless he does enough to acquire the fundamentals, his achievement is minimal. (Consider learning a language: Without a basic grasp of the grammar, individual bits of vocabulary mean nothing.) Once these fundamentals are present, however, new possibilities open up. A person can use his basic concepts and skills to acquire further ones, and for a time his progress is rapid. Because of this, diversity of pursuits is most damaging when it prevents a person from acquiring any fundamentals. If he never reaches a takeoff point, he has no genuine achievements. But the same diversity is less harmful further along. Someone with the elements of several perfections can make real gains in them all and need not be distracted by variety. Where a beginner must concentrate, those who are more advanced can alternate among activities, returning to each with their mastery secure.

The costs of concentration, by contrast, are greatest at high levels. Someone who devotes an hour a day to history is in little danger of fatigue and can easily add an extra hour. Doing so may even increase his commitment. But after eight hours boredom and fatigue are real factors, and it is questionable how much someone working this hard can add in an extra hour per day. It might even improve his history to relax by doing something different. What is more, diminishing returns should affect perfections mainly at high levels, which does seem plausible. In many domains, the rapid progress that follows acquisition of the fundamentals eventually slows. As a person approaches the frontiers of a discipline, further advances become more difficult, requiring more effort and application. The return of perfection to time becomes less, not more.[19] Finally, there are interaction effects. Although diversity of pursuits is distracting at low levels, it can be fruitful further on. Major intellectual advances often occur when insights from one area are applied to another, and those who seek such advances will do well to broaden their knowledge.

These arguments suggest a unified picture: Dilettante's disadvantage is strongest at low levels of perfection and the costs of concentration are greatest higher up. This in turn suggests that many people's achievement lines may be M-shaped, as in Figure 7.4. In the upper left and lower right of Figure 7.4, where the achievement of one good is high, the costs of concentration work strongly on that good to make the

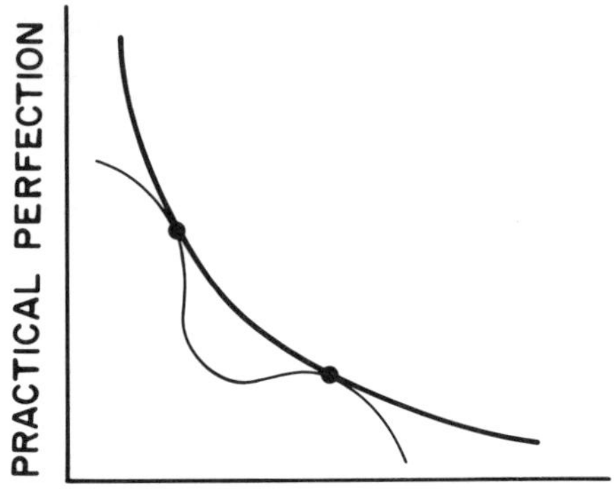

Figure 7.4

line concave. In the centre, however, where both perfections are low, the disadvantage dominates and the line is convex. If the M-shape characterizes a person's options he has two better lives available, each concentrating to some extent on one good. He must be neither a pure specialist nor a pure all-rounder, but give a moderate preponderance to one perfection. To specialize more than this is to fall afoul of the costs, which is serious. But it is also wrong to aim at too much balance. Someone who does that will slide into the central trough of his achievement line and onto a lower curve.

In Figure 7.4 the M-shape dominates, and the moral upshot would not be very different given constant trade-offs. But balancing still plays a role. For a start, it makes the ideal life slightly more balanced, with its point on the inside slope of a hump rather than right on top. So a person's goal is slightly more proportion. It also affects the badness of less than ideal lives. Because the indifference curves dip into the central trough, a dilettantish life is, although not best, less bad than if the curves were straight. And an overly specialized life is worse. Even with M-lines, balancing affects deliberation, pushing us towards proportion and away from monomania.

A person with two ideal-life points, as in Figure 7.4, can choose between two equally good lives and do so as she pleases. Given precise cardinal comparison, such exact equality would be unlikely because most people's abilities incline them somewhat more to one perfection. But, given the merely rough comparability of perfectionist values (7.1.3), something like it may be common. Many people may have several good lives available, each concentrating moderately on one good, and be morally free to choose among them. Having made an initial decision they must stick to it, for otherwise they will lapse into dilettantism. Yet, at the start their choice of lives is morally unconstrained.

7.3.4

If I am right about the shape of (many) achievement lines, they answer a common objection against balancing, namely, that it promotes mediocrity. Rashdall urges this objection against a "self-realization" theory that identifies the good with "an equal, all-round development of one's whole nature":

> Up to a certain point the man who is a mere specialist will be a bad specialist, but that point is soon reached. Charles Darwin found that the cultivation of reasoning power and observation had extinguished his once keen imagination and sensibility. And yet who would wish—whether in the interests of the world or in the interests of what was best in Charles Darwin's own nature—that his work should have been spoiled in order that one of the three hours which was the maximum working day his health allowed should have been absorbed by politics or philanthropy? Who would decide that the origin of species should have been undiscovered, in order that the man who might have discovered it should retain the power of enjoying Wordsworth? This notion of an equal, all-round, "harmonious" development is thus a sheer impossibility, excluded by the very constitution of human nature, and incompatible with the welfare of society. And, in so far as some approximation to such an ideal of life is possible, it involves a very apotheosis of mediocrity, ineffectiveness, dilettantism.[20]

This entertaining passage exaggerates its history. Darwin discovered the theory of evolution some twenty years before publishing *The Origin of Species* and continued scientific work for twenty years after publication of that work. Therefore, a realistic picture of his life without an hour's biology a day will have it lack, not the discoveries that made him famous, but perhaps the works on *Insectivorous Plants* and *The Effects of Cross- and Self-Fertilization* that he published late in his career. Is it wrong to say that his life would have been more complete had he left the work of these books to others and done more outside biology? I do not think so, and neither did Darwin. In a memoir written by his son, he is reported to have told one of his daughters that "if he had to live his life over again he would make it a rule to let no day pass without reading a few lines of poetry. Then he quietly added that he wished he had not 'let his mind go to rot so'."[21]

The Darwin example may be used to make a stronger claim than Rashdall's. Some may say that if Darwin's greatest talent was for science, he should have preferred even tiny achievements there to large ones in other areas. He was above all a biologist, and in judgements about his life his biological perfections count above all others. This view is not the lexical view discussed above (7.1.1). It does not say that one perfection is higher in all humans, regardless of their talents or situation. But it still supports specializing conclusions. It says that, although different perfections are in themselves equally good, the value of a person's life depends primarily on what he achieves in his single best one. If his talents incline him more to one good than to others, he should prefer any gain in it to improving them.

This *specializing* view may be attractive to some, but it directly opposes the balancing view and flouts the intuitions I most want to capture. In my view the most appealing ideal, especially when we consider the value in whole lives, is that of well-roundedness. And we can retain this ideal if the most common objection to it, Rashdall's objection, is answered by M-lines. Because these lines make too much balance undesirable, a view that values roundedness can agree that if appreciating Wordsworth had prevented Darwin's major discoveries, it would have been wrong. At the same time, it can say that of the broadly specialist lives available to

Darwin, the best contained at least some other goods and were not solely devoted to biology.

The combination of balancing and M-lines yields a perfectionism in harmony with many intuitions about specialization. On the one hand, it makes roundedness an ideal and favours lives with the greatest balance consistent with serious achievement. On the other hand, it reflects our suspicion of dilettantes. The combination may also help historically. When classical writers discuss theory and practice, they often compare, not knowledge and action as such, but whole lives devoted to them. They ask not whether understanding is better than political rule (for many the highest practical activity), but whether a philosophical or contemplative life is better than a political life. This makes sense on the view we have described. It follows from balancing that we cannot compare two excellences directly. Because a unit of theory that would mean a great deal to a politician will add little to a scholar, we cannot evaluate it apart from the life in which it figures. The smallest unit we can compare is the complete human life. If this explains the focus on lives, the restriction to specialist lives may follow from M-lines. If we want to compare the best lives commonly available, dilettante's disadvantage tells us to ignore any that divide their time equally. We can consider contemplative and political lives, but not ones that aim at both goods. One perfectionist who does discuss balanced lives is Plato, in his account of the philosopher-kings. But he insists that these kings, precisely because they combine such different abilities, will be very hard to find.[22]

7.3.5

A final point is this. M-lines are plausible because dilettante's disadvantage is strongest at low levels and the costs of concentration higher up. If this applies along one achievement line, it should also hold for a series of lines representing the options available to different people. Those with limited abilities or a short lifespan should find that dilettante's disadvantage works powerfully on their options, creating a deep central trough in their line. For those with more talent and time, however, this factor should be less important. As their achievement lines move further from the origin, the troughs should become shallower and the line eventually approximate the simple concavity that would obtain given only costs of concentration (Figure 7.5). This too matches intuition. It implies that, although exceptional individuals like Marcus

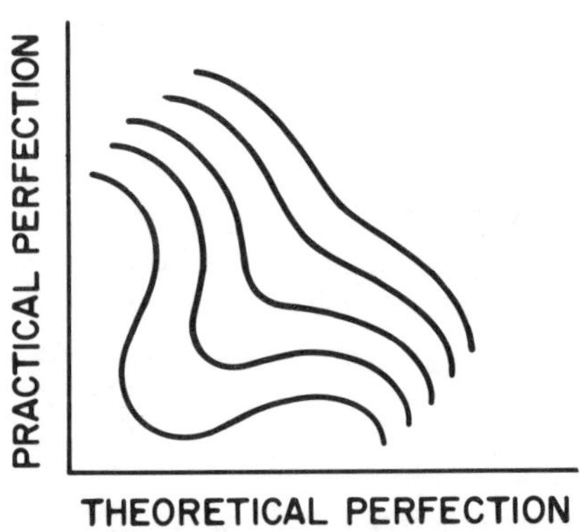

Figure 7.5

Aurelius and Leonardo can aim at well-rounded lives, those with more limited gifts must content themselves with being some kind of specialist.

7.4 Many-Person Balancing?

7.4.1

So far we have applied balancing within lives and made the relative weight of a perfection depend on the relative degree to which a person herself has achieved it. But it is also possible to balance across lives. We can define an ideal proportion for a collective and say a group does better when its members achieve a variety of goods than when they all specialize alike. Just as we value roundedness within lives, so we can value it across them, and prefer groups whose members exercise a diversity of talents.

For some perfectionists this many-person balancing is a simple extension of the one-person view. In his account of "harmonious development," Humboldt says that what occurs in one person "by the union of the past and future with the present, is produced in society by the mutual co-operation of its members." An individual can achieve only some excellences open to the species, but by entering into a "social union" with others she can overcome this limitation and "participate in the rich collective resources of all."[23] The intuitive idea here is that, if she belongs to a group whose members achieve all the goods, she can participate through them in a well-roundedness denied her on her own. But the idea again rests on balancing. A group can offer this participation only if some members' achieving a good permits others to neglect it and gives them a reason to seek other perfections instead.

Far from extending one-person balancing, however, this idea can conflict with it. Imagine that a person has achieved more theoretical than practical perfection but belongs to a group whose other members have preferred practice. Balance in his life requires more action, while proportion in the group demands the opposite. The same conflict is possible for different groups. A person's family may have achieved more theory than practice, but his nation the reverse. Which should he prefer? Balancing applied within different units makes different demands, and a morality that extends it faces difficult choices about where its claims take precedence.

Because of this, a perfectionism with many-person balancing will be very complex, with a different overall goal for each collection where proportion is valued. This is hardly a decisive objection, but I think the best perfectionism should confine its balancing to single lives.

7.4.2

We should agree first that the intuitive appeal of balancing is greatest in single lives. Here an ideal of proportion strikes us immediately, whereas for larger groups it is less obvious. There is a reason for this: The degree to which balancing is plausible for a collection depends on the degree to which the collection forms a unity. The states making up a single human life are connected in especially intimate ways and form a tighter unity than any collection across lives. This unity makes it especially natural to consider them together and to seek proportion among them. As confirmation, note that when we do balance in groups it is only the most unified that we consider. We may praise a family if its members achieve different excellences, or admire a civilization like Greece for excelling in such diverse areas as poetry, warfare, and philosophy. But if we consider accidental collections like the group of people born on a Tuesday or those with names containing seven letters, the same approach is not plausible.

This difference suggests that decisions about balancing reflect ideas about where significant unities lie, and here narrow perfectionism has decided views. Its ideal of developed human nature is achieved by humans living lives, and in elaborating the ideal we have tried to emphasize lives as morally important units. We have made perfectionism time-neutral, so agents have one goal at all times (5.2.2), and required it to aggregate first across times in a life (6.1–6.3). We carry this tendency through if we now prefer one-person balancing. As we have seen, this balancing prevents us from comparing two excellences directly (7.3.4). An item of knowledge that would be good in a politician's life may add little value to a scholar's, and we cannot assess it apart from the life in which it will figure. Perfectionist judgements presuppose comparison, and given one-person balancing, comparison requires first assigning goods to lives. The same is not true of many-person balancing, which can compare once it knows the total in a group. So the many-person view gives no special place to lives. It would respect perfectionism's tendencies a little if we balanced everywhere, but gave the consequences in single lives the greatest moral weight. But it is surely simplest to stick to lives. This restricts us to the balancing with the greatest intuitive appeal and stays closest to our original perfectionist ideal.

8

Trying, Deserving, Succeeding

We now turn to the first part of perfectionist measurement, the account of the individual perfections on their own. As before, we cannot expect precise cardinal measures, but ask only what counts for and against the perfections and by roughly how much. If we consider mathematically precise principles it is only hypothetically, and assuming supervaluations across them (7.1.3).

In introducing physical perfection, I said its measurement is more a subject for physiologists than for philosophers (4.1.2). So I will set it aside here as largely beyond our competence. Instead, the next three chapters examine theoretical and practical perfection. They ask what states contribute most to these perfections or most develop theoretical and practical rationality.

An account of these perfections must observe the constraints imposed by the concept of rationality as that is essential to humans. It must define degrees of perfection using only criteria this concept allows, and none it forbids. These constraints do not single out unique measures, but leave some play room for moral choice (2.3.3). I am uncertain, however, of the extent of this play room. How far does the concept of rationality constrain our notions of excellence? How many alternative measures does it allow? In what follows I will often present different accounts of theoretical and practical perfection and, unable to choose one on conceptual grounds, will leave the issue to moral intuition. If one is preferred, it will be because it better matches our moral judgements. A more sophisticated approach might narrow the play room here and derive a more determinate picture of excellence from the bare idea of developing rationality. If so, this approach would mark an advance, and just one way in which what I offer as tentative initial proposals might be improved.

8.1 Number and Quality

8.1.1

To measure the rational perfections, we first need a general framework. Ours will measure along two dimensions, *number* and *quality*. A person's perfection will depend both on the number of states of some kind she has and on their standing on a scale of quality. This framework has two virtues.

First, it is intuitively appealing. To see this, consider the common view that equates theoretical perfection with knowledge or, more specifically, with justified true belief. (There are alternatives to this view, but it illustrates a general point.) Its most plausible versions make each person's perfection depend partly on the number of truths she knows. If she acquires a new justified true belief, this increases the value in her life; if she loses one, her theoretical good diminishes. But number cannot be the only test, for plainly some beliefs are more worth having than others. It is better to know a fundamental law of the universe than the number of redheads in Beiseker, Alberta, or the workings of a friend's personality than the exact length of his forearm. On the most attractive view, the value of someone's knowledge depends on two factors: how many truths she knows, and their quality or importance.

The same structure is appealing for judgements about practice. Parallel to the view that values knowledge is one equating practical perfection with the successful achievement of one's goals, given a justified belief that this success would happen. Here too a person's good is partly a matter of number, this time of the number of his non-lucky achievements. But there are also considerations of quality. Someone who through elaborate planning achieves a major reform of his society does more of intrinsic worth than someone who merely ties a shoelace. Even apart from its benefits to others, his act does more to make his own life excellent. On the practical side, too, a person can increase perfection either by increasing the number of his deservedly successful goals or by increasing their importance.

Beyond its intuitive appeal, the framework fits our Aristotelian theory of human nature. According to this theory, humans are rational because they can form and act on sophisticated mental states, ones with extended contents and complex hierarchical relations (4.2). This suggests that there are two ways humans can develop rationality: by having more of the relevant states or by having states that are more sophisticated. These are precisely the options represented by number and quality.

Given this dual support, the number-quality framework will guide our discussion of measurement. This chapter discusses the dimension of number and the different views possible about the kinds of state whose presence adds to excellence. The next two chapters then examine quality. More specifically, they develop an account of quality that is recognizably Aristotelian and uses the same tools to judge theory and practice.

8.1.2

The number-quality framework is modelled on Bentham's account of hedonic values. He, too, thinks it better to have more numerous value-bearing states—on his view more pleasures—and also ones of greater intensity.[1] But there is an important difference between the models.

Bentham's theory recognizes not just an intrinsic good, pleasure, but also an intrinsic evil—pain. Just as more intense pleasures are better, so more intense pains are worse. Narrow perfectionism, by contrast, does not permit talk of intrinsic evils. Because an essential property cannot be realized to negative degrees, the theory's scale of quality must have zero as its lowest point. It must say that every state that passes the tests of number has positive perfectionist value, and that, considered on its own, every human life is worth living. This implication was recognized by several perfectionists and appears most clearly in the claim of Aquinas and Leibniz that evil is nothing real, but only the absence of good.[2] This claim need not deny that pain and moral vice, for example, are real, or that there are states whose causal properties make their existing on balance regrettable. It just says that no human states have intrinsically negative value, which must be true if the ground of value is the development of essential properties.

Intrinsic evils can be accommodated within a theory that is not just narrowly perfectionist. A merely broad perfectionism can treat moral vice or false belief as intrinsically evil (see further 10.4.1); a theory that gives some weight to hedonic values can treat pain as intrinsically evil. Some may think their recognition of evils a reason to prefer such theories, but in pure narrow perfectionism there can be only positive values.

8.1.3

The moral issues about number concern the conditions a state must meet before it and its quality add to excellence. The main debates here assume agreement on some general categories in which perfections occur. For theoretical perfection I will take this category to be that of *beliefs*. A person's theoretical good at a time will depend on the number and quality of (some of) the propositions she believes at that time, so the issue is whether all her beliefs count or only those that are, say, justified or true (8.2–8.4). On the practical side, the general category will be *intentions*. The relevant facts will concern the ends a person intends at each time or has resolved actively to pursue.

What is more, I will interpret these categories *dispositionally*, so what count are the propositions a person is disposed to assent to and the ends she has a dispositional resolve to pursue. I do so partly to meet a requirement of narrow perfectionism. If rationality is essential to humans, it must be realized to some degree at every time in their lives, and, given the number-quality framework, some states are required at every time. Because dispositional beliefs and intentions persist through sleep and unconsciousness, they help satisfy this condition.

Some perfectionists reject this dispositional view. Aristotle and Aquinas, for example, think that theoretical perfection, at least in its highest form, comes not in having but in contemplating beliefs.[3] (A parallel view values most the ends a person consciously intends at each time or, perhaps, is actively pursuing.) Their motive for this view is partly to make perfection more active. As Aquinas argues, our highest good should be not a "potentiality" but an "operation," and "he who knows is potentially considering."[4] Contemplation, in other words, realizes the potential in dispositional belief.

Despite the appeal of this argument, I do not believe the contemplation view has a place in the best perfectionism. In its pure form, the view does not allow states at every time, and it also has counter-intuitive consequences. Imagine that one person knows one truth, which he contemplates at every time in his life, whereas another knows many truths that she contemplates in succession, one after another. On a pure contemplation view, the first person has, quality aside, the same theoretical perfection at each time as the second and, given standard aggregative principles, the same perfection in his life as a whole. This is surely implausible.

A weaker view gives just more weight to contemplated than to dispositional beliefs, but it, too, does not fit our intuitions about theoretical value. Do we think physicists should take time off research to contemplate the laws of quantum theory, or historians to mentally enumerate the events of the French Revolution? Surely not. What matters intuitively for theoretical perfection is how, dispositionally, a person pictures or understands the world; it is what beliefs she has available to contemplate if she chooses to. The contemplation view might be reasonable if there existed a God or other supreme being, for then contemplation might be a proper response to his goodness or show him a proper obeisance. Without such a being, however, the bearers of theoretical value should be some dispositional beliefs.[5]

A different view equates theoretical perfection with the search for beliefs. Lessing says that if God offered him a choice between Truth and the Search for Truth, he would unhesitatingly choose the Search for Truth. Malebranche says, "If I held truth captive in my hand I should open my hand and let it fly, in order that I might again pursue and capture it."[6] Reflecting these claims, a search view of number counts, not any beliefs a person has at a time, but those she is actively pursuing.

Although this view is appealing we need not accept it, because we can capture the intuitions behind it while still using beliefs as our general theoretical category. We can, first, value the search for beliefs instrumentally. Through active research, people make new beliefs possible for themselves and others and also deepen their understanding of their existing beliefs. Their seeking, therefore, is often good as a means. Further, we can value the search for beliefs intrinsically, but as an instance of practical rather than theoretical perfection. This is, in fact, the proper way to value it. What, after all, does a scientific researcher do? She sets herself a goal—discovering a law or explaining some phenomenon—and uses skill and ingenuity to achieve it. In carrying through her research plan she uses the same rationality as an engineer or politician. She pursues some goals—data collection, a crucial experiment—as means to others and so works towards a final result. Although her subject is intellectual, her special excellence is to achieve ends and is therefore practical. More specifically, it combines extensive prior knowledge with active problem-solving to expand that knowledge, uniting theory and practice in a way that, given balancing, is especially valuable (7.2.3). Once we see that the value of research need not be theoretical, we can accommodate it while still using beliefs as a general theoretical category.

A rival view of practical perfection counts, not a person's dispositional intentions at a time, but her dispositional desires. This view, however, ignores the central role of intentions in practical reasoning (4.2.2) and also makes practical perfection less active than it can and should be. It is counter-intuitive to let a person's excellence be affected by the ends she idly wishes for, as opposed to those she has set herself to pursue, and we avoid this result by having number count only intentions.

8.1.4

Especially given these general categories, the number-quality framework may seem vulnerable to objections. It may be argued, for example, that there is no uniquely correct individuation of a person's dispositional states—no one way of counting beliefs and ends. Even if this is so, however, we can make the rough count of states that results from supervaluations over a set of acceptable individuations, and the rough value-judgements this kind of count allows. It may also be argued that, even if there were one correct individuation, the resulting number of states would always be the same: infinitely many. Any time a person dispositionally believes or intends *p*, she does the same for *p & p*, *p & p & p*, and so on to infinity. This last cannot be a decisive objection. We understand informally how one person can know more truths in a given area than another, or more than before; and several devices can capture this understanding. The one I prefer has the dimension of number count, not individual dispositional states, but the number of states *of different kinds* a person has. We imagine a division of truths or ends into finitely many kinds, so *p*, *p & p*, and *p & p & p* are in one kind, and make a person's excellence depend on the number of kinds in which she has relevant states.[7] This device ensures finite measures of number, but in so doing makes them even less precise. Because there is no uniquely correct division of states into kinds, we must supervalue not only across acceptable individuations but also across acceptable divisions. The result can be only very rough cardinality.

8.2 Attempt

With these general categories fixed, we can address the main debates about number. These concern the choice between four views that I call *attempt, deserving attempt, success,* and *deserved success*. Given precise cardinality, these views would say a person has one state of theoretical perfection for every

1. belief he has (attempt);
2. justified belief he has (deserving attempt);
3. true belief he has (success); and
4. justified true belief, that is, item of knowledge, he has (deserved success).

For practical perfection, they would find one state for every

1. end he intends in the belief that he will achieve it (attempt);
2. end he intends in the justified belief that he will achieve it (deserving attempt);
3. end he intends in the true belief that he will achieve it, that is, end he successfully achieves (success); and
4. end he intends in the justified true belief that he will achieve it, that is, end he successfully and non-luckily achieves (deserved success).

Variations on these views are possible. For example, a mixed view gives some weight to mere belief, more weight to justified belief, and even more to true belief. But to set out the moral issues here, we should start by examining the simple views. And in doing so we should assume that our perfectionism will use the same view for both theoretical and practical perfection. Because both develop rationality, it is natural to characterize them in parallel, using the same measures of number and quality. So let us consider the four views in turn, beginning with the least demanding, attempt. And let us assume throughout a constant account of quality, that is, a constant account of the comparative worth of the states that pass the different tests of number.

8.2.1

Attempt is very generous about number, letting every sincere belief and intention bear value, regardless of their truth, success, or justification. Because of this, attempt leaves the bulk of perfectionist evaluation to quality. With no other conditions to prevent a dispositional state from having value, people can and should increase their states' quality without limit. The contrast with the other views is dramatic.

In their various ways, the three more restrictive views of number reflect a *pragmatic* conception of practical rationality. On this view, rationality requires us to tailor our ends to our abilities. The rational agent forms an accurate picture of his talents and situation and pursues only ends that they make realistically attainable. Otherwise, his activities will not be successful, or if they are they will not involve a justified belief. By contrast, attempt expresses a *romantic* view of rationality. It cares only about the fineness or sophistication of our ends—those factors captured on the dimension of quality—and not at all about our success or likelihood of success in achieving them. It prefers a vain pursuit of noble goals to success at something mundane. Whereas pragmatists tell us to restrain our ambitions, romantics want them to run free. We should aim at the stars, not to achieve more on earth, but to have our sights on what is noblest in itself.

Romanticism is also possible about theory. Here the more restrictive views reflect a pragmatic demand to adopt only beliefs that we have good reason to believe are true. These beliefs are grounded in evidence and are unlikely to mislead. Attempt, however, makes our theoretical perfection depend solely on the scope and explanatory integration of our beliefs—factors captured by quality—and not on anything connected with their truth. Whereas pragmatists want our beliefs restrained by evidence, romantics tell us to soar beyond evidence. What matters for them is not how well our theories do or should match the world, but how much the world would be worth knowing about if they did.

The romantic view has many literary expressions. Browning's Bishop Blougram says, "'But try,' you urge, 'the trying shall suffice: / The aim, if reached or not, makes great the life.'" Then there is the famous "Ah, but a man's reach should exceed his grasp, / Or what's a Heaven for?"[8] But the best illustration of romanticism is Don Quixote. Inspired by a chivalry that no longer fits his world, Quixote pursues beautiful goals that cannot be realized. Although his ends are admirable, in his surroundings they cannot be accomplished. For pragmatism the impracticality of Quixote's ideal makes him a foolish, even pathetic figure; for romanticism it is a sign of nobility. When the world no longer permits honourable achievements, one's best response is to seek finer goals beyond it. Note, though, how far-reaching the response must be. With intentions as our general practical category, Quixote must actively pursue his ideal. If he stayed at home merely wishing to act chivalrously, he would be no subject for a novel. Because of this, Quixote must believe against evidence that his ends are achievable: that windmills are giants and a scullery maid a lady. Pragmatism finds the irrationality of these beliefs a further aspect of Quixote's foolishness, but for romanticism it adds to his perfection. When the world no longer contains truths worth believing, one must find beautiful falsehoods and attach oneself to them.

8.2.2

In rejecting pragmatism, attempt also rejects a related implication of the more restrictive views. They require the tailoring of ends to abilities because this tailoring makes successful achievement more likely. But it also makes achievement more likely if one chooses effective means to goals once these are present, and the restrictive views therefore require this choice. They demand skill at deliberation and execution, at identifying and achieving the intermediate goals that will accomplish one's final purposes. Attempt disagrees. It acknowledges that pursuing an end requires a belief that one's means will be effective, but whereas the other views insist that this belief be justified or true, attempt is satisfied with its presence. (Forming the belief may require wishful thinking, but attempt allows wishful thinking.) The fact that our routes to valuable ends are not the best is immaterial if the ends themselves are best and our pursuit of them is sincere.

What attempt rejects, at all these points, is a general view about prudence. Tailoring ends and choosing effective means are aspects of the traditional ideal of prudence, and views that require them treat prudence as a moral virtue. They hold that skill with one's ends is needed for morally valuable conduct. This moralizing view of prudence is striking, and its possibility, given anything other than attempt, distinguishes perfectionism from that rival tradition in ethics that tries to reduce morality to prudence. Where writers like Hobbes hold that morality as commonly understood is a requirement of prudence, restrictive versions of perfectionism make prudence just one part of the life morality finds desirable.

This moralizing of prudence is common in the tradition. Aristotle, for example, thinks that to perform the best acts we must not only aim at the right mark but also take the right means, where knowing these is the job of prudence.[9] His view is, in fact, the standard one. Most perfectionists want us to base our beliefs on evidence and to pursue ends in ways that make their achievement likely. But the three views that support these claims derive them from different conditions, with different implications for other issues. Assuming pragmatism about ends and a demand for effective means, let us now see what follows from the different restrictions imposed by deserving attempt, success, and deserved success.

8.3 Deserving Attempt

To the generosity of attempt, deserving attempt adds one condition: that our beliefs, about the world or the future success of our actions, be justified. What is justification? For our purposes justification is best read on an "internalist" model, so a belief's justification depends solely on factors within a person's mind. More specifically, it is best read so a belief's justification depends on its probability given all a person's other beliefs, discounted by the initial probabilities of some of those other beliefs. The idea here is that justification requires coherence. If a belief meshes with other beliefs, it is or may be justified; if they make it improbable, it is not. The rival "externalist" model makes justification depend on a belief's causal history, especially on the reliability of the mechanisms by which it was acquired. This model is less traditional and happens to have less interesting consequences for ethics. So we will understand justification internally, as involving a coherence with other beliefs reflected in high conditional probabilities.

8.3.1

One effect of requiring justification is to abandon romanticism. If only well-grounded beliefs count, we cannot increase our excellence by believing against evidence or pursuing unrealistic goals. With only justification required, our beliefs need not be true or our projects successful; in this respect the condition is weaker than it might be (8.4). But in another respect it is stronger.

On deserving attempt and deserved success, we must be prudent, not only globally, but case by case. We must accompany each of our goals by a justified belief that we will achieve *it*. This is not so on the other one-condition view, success. According to success, we must always discount the probability of our not achieving an unlikely goal. If we somehow do achieve it, however, we will get full credit for all its quality. So if the goal's quality is sufficiently high, choosing to pursue it may be antecedently right; it may be worth risking probable failure for a chance at glorious success. This cannot be so given a justification condition. Then achieving an antecedently unlikely goal, even should it happen, will have no value and pursuing it cannot be right. On the desert views, prior probability of success is not just one factor to be weighed against others, but a precondition of valuable activity.

8.3.2

Deserving attempt also grounds, and unites with other claims, the idea that rationality involves consistency in one's beliefs and ends. Other views can capture this idea by stipulating that neither of two inconsistent states counts for excellence. (I imagine attempt and success doing this.) But a desert view entails it.

A person who has contradictory beliefs, believing *p* and *not-p,* has two beliefs that do not cohere with each other. The probability of either belief given *all* her other beliefs is therefore zero, and each is entirely unjustified. If she has the first, she ought not to have the second, and vice versa. Deserving attempt therefore entails what other views have to stipulate: that neither of two contradictory states has value. And it goes further. Imagine that someone believes *p* on evidence *q* and *r,* but also believes *not-p*. If the dimension of number refused only to count *p* and *not-p*, it would judge her as no less rational than someone who believed *q* and *r* but was agnostic about *p*. She would do no worse believing a contradiction than if she merely failed to draw an inductive consequence. This implication is counter-intuitive, and deserving attempt can make a stronger claim. If *q* and *r* support *p,* then not only is *not-p* improbable given them, but they are improbable given it. *Not-p*'s failing to cohere with other beliefs makes them not cohere with it. Deserving attempt therefore excludes as bearers of value not just contradictory beliefs but also any further beliefs that are evidence for them. The rot of unreason spreads, making additional beliefs unjustified that would count for value if the person merely failed to conclude from evidence.[10]

Similar points apply to consistency among ends. Because intention implies belief, someone who intends *p* and *not-p* believes that she will achieve *p* and *not-p*, and given the other belief, neither is justified. The same holds if she intends *p* and some other end that she has evidence is contingently incompatible with it. Again the rot of unreason can spread. After intending *p*, she may intend *q* and *r* as means to it. Because achieving *q* and *r* precludes achieving *not-p*, her beliefs about *q* and *r* are improbable given her intention that *not-p*. Incompatible goals cancel not just each other but also any further goals that are means to them.

Note that practical inconsistency involves conflicting intentions, not just conflicting desires. It is not irrational to desire *p* for one reason and *not-p* for another, and to have difficulty choosing between them. Irrationality enters only if one goes beyond this to form conflicting resolutions, that is, simultaneous intentions to achieve two incompatible ends. Deserving attempt condemns only the latter.

To illustrate these points, consider a particular practical inconsistency.[11] In modern societies, many women are brought up to have conflicting long-term aims. On the one hand, they are raised to value gentleness and a willingness to help others; on the other, they are encouraged to seek success as our societies define it. But this success involves achievements that are either inherently competitive or open only to those who can aggressively do down others. These women therefore enter adulthood with contradictory ambitions. Not only do they desire both to be gentle and to be successful, which might be harmless, but they intend both. This could diminish their perfection even without a justification view of number. People whose goals conflict often oscillate between them, with their efforts under one head undermining their efforts under the other. On any success view, this oscillation deprives them of excellence, but a desert condition makes the loss greater. Then the rot of unreason spreads, cancelling other goals that are means to the two in tension. Not even successful bits of modesty or careerism add to their perfection. When irrationality infects a person's leading projects, whole stretches of action lose value.

8.3.3

As part of its claims about consistency, deserving attempt can condemn weakness of will, or failing to act as one thinks all-things-considered best. Weakness of will is often considered a perfectionist evil. Aristotle thinks it is incompatible with the full development of practical rationality, and Nietzsche makes it his chief imperfection.[12] As we will see, much of their view is captured on the dimension of quality: Weak agents act on less valuable intentions and also diminish the worth of their ends at other times (9.4.3). But their loss would be greater if they also had conflicting intentions, intending both to act as they think best and, simultaneously, to do the weak thing they do. There is the following to be said for this analysis: Before a person acts weakly, a contrary intention is often clearly present. Before breaking a diet, for example, he may intend to stick to it, and this intention may persist to a time right before he lapses. Given the prior existence of this intention, it is simplest to explain weakness as occurring when another intention intervenes and prevents the first from having its normal effect. But then weakness is an even greater imperfection. The weak agent loses not only the quality in his better intention, which he is prevented from acting upon, but also the lesser quality in the one that causes his action. With rot-spreading, the loss may extend to other ends that are means to the one that causes his action. Like other inconsistencies, weakness creates a hole in the system of ends, negating any quality there.

If we value rationality, we will, minimally, require agents to have consistent beliefs and intentions. A desert view entails a strong version of this requirement and unites it with other demands, for example, for case-by-case prudence.

8.4 Success and Deserved Success

There remain two final views, success and deserved success, with their shared demand for true beliefs and successful intentions. On its own, in the success view, this demand supports pragmatism (although only globally) and the need for effective means. Yet, what does it add when these implications already follow from a justification condition? What is its distinctive contribution to the most restrictive view, deserved success?

8.4.1

A success condition alters the general character of the good. On the attempt view, a person's excellence depends solely on states within her mind, and the same is true, given internalism about justification (8.3), for deserving attempt. If quality too is internal (chapters 9–10), we can assess her excellence without looking outside her mental states. The success views, by contrast, make perfection partly relational. On the theoretical side, they require a relation of correspondence between a person's beliefs and the world, and for practice a similar correspondence between the world and her intentions. In the one case, her beliefs must match what already exists outside them; in the other, the world must come to match her goals. In both cases, however, some matching is vital. Although other aspects of perfection, such as quality and justification (if required), are internal, they count for nothing without the right external relations.

There is a general issue here. Theories of the good can find intrinsic value in three places: in states of the world apart from relations to a mind, in states of a mind apart from relations to the world, or in the obtaining of relations between a mind and the world.[13] In its account of rationality, perfectionism cannot take the first line, but its view of number decides between the other two. Whereas attempt and deserving attempt keep excellence within the mind, success and deserved success require relations beyond it.

Although clear enough abstractly, this issue can seem remote from practical deliberation. How can we aim at truth except by forming beliefs that are justified or achieve our ends except by taking means our evidence says will be effective? The issue can bear on deliberation, most clearly when our acts affect others. I can know that another person has beliefs that, although consistent with her evidence, are false, or is pursuing an end by means that will not in fact achieve it. On the attempt views, these facts give me no reason to correct her beliefs or advise a change of plan. On the success views, by contrast, they do. A similar situation can arise within oneself. I can know that if I act some way now I will later have beliefs that, although then justified, are false. The attempt views may recommend this action, but the success views do not. Finally, even when the issue does not bear on deliberation, it can affect our attitudes. The good is what we ought to desire (5.1.2), and on the different views of number what we ought to desire is different. Imagine learning that an important belief one held for many years was false or that a project long pursued will now end in failure. According to deserving attempt, this discovery need be no cause for regret. If one's beliefs in the past were justified, they had all that matters for excellence. On the success views, however, learning that one's past states lacked relations to the world should cause one chagrin.

A dramatic illustration of this issue is provided by the fantasy of an "experience machine."[14] This device, by electrically stimulating the brain, gives the illusion and thus the experience of any desired activity. If this machine existed, would using it be right? If it does not exist, should we rue its absence?

Hedonism notoriously answers these questions in the affirmative, but so do some perfectionist views. Attempt, for example, has no objection to the experience machine. If our states on the machine have sufficient quality—and this can be programmed in advance—attempt permits and even encourages its use. The same goes for deserving attempt. Given our experiences on the machine, our beliefs about our status and the goals we are achieving are perfectly justified: They do not contradict but follow from our evidence. The success views, however, find the machine repellent. For them, the experiences it provides are impoverished because they lack connections to outside reality. People who plug in do not have true beliefs about their surroundings and never actually achieve the ends they intend. They are severed from reality, and in becoming severed have chosen something wrong.

8.4.2

Closely tied to the issue of relations are questions about the moral significance of luck. By keeping perfection within the mind, the attempt views make it largely proof against luck: If people can control their mental states, they can provide sufficient conditions for their excellence.[15] On the success views, however, perfection is partly dependent on luck.

This is most thoroughly so on simple success. This view allows a person's good to be both increased by undeserved good luck and decreased by bad. Unjustified beliefs that turn out to be true—think of lucky guessers in science—or ill-made plans that succeed increase a person's perfection. And perfection is diminished if states that deserve to be related are not. Deserved success rejects the first of these claims: Because it requires justified beliefs, it gives no weight to undeserved good luck. But it does count undeserved bad luck. Justified beliefs that turn out false or well-planned activities that fail deprive people's lives of value. If through no fault of theirs states that should match reality do not, their excellence is lessened.

This last claim may be resisted. Consider a great scientist who for all his brilliance in research ended up with beliefs that are largely false. Consider, for instance, Aristotle in his career as a biologist. Because most of Aristotle's beliefs about biology were false, a success view must say he achieved little theoretical perfection—less, probably, than an undergraduate biology student today. Surely, a defender of deserving attempt may argue, this claim is implausible. Perfection should be more self-sufficient than this (5.1.3), as it will be if it requires only justified beliefs. Then those who respond well to their evidential situation, as Aristotle did, achieve high theoretical perfection.

This critique may be taken further. Not only is it attractive to make perfection self-sufficient, some may argue, but it is also a requirement of narrow perfectionism. If the rational perfections develop aspects of human nature, they must occur within human lives, and this excludes the relational views. On success and deserved success, the bearers of value are not individual humans but complexes involving a human and the external objects she knows or brings about.

This last argument is not persuasive. We can understand a person a's knowledge of b as involving a two-place property ... K ... true of a complex $\langle a,b \rangle$. But we can equally well understand it as attaching the one-place property ... Kb to a, so it is true of a that she knows b. And there are positive reasons to understand it this way. Alongside truth, knowledge involves justification, which on our view is internal. So the state we value has an important purely mental component, and the reason this supplies for treating knowledge as a state of humans is strengthened if quality, too, is internal (chapters 9–10). When a partly relational state is largely internal, there are positive reasons for ascribing it to the knowing mind.

The success views may attempt their own stronger claim: that narrow perfectionism requires the good to be relational. As humans, they may say, we live within a world, affecting and affected by it. Our lives are essentially situated, and accounts of our perfection must reflect this fact. Instead of abstracting from relations to an environment, they must give them positive weight.[16] I am uncertain whether this argument succeeds (see 8.5). If it does, a relational view of number is required by facts about human nature. If not, we have a moral choice between two ideals: of external connection and, opposing it, of personal self-sufficiency.

8.4.3

Because they value relations, the success views make distinctive claims about time. On most theories of the good, a state's value now can depend only on facts obtaining now, but in a relational perfectionism this is not so. Imagine that a person believes some event will occur in the future. Whether this belief amounts to knowledge depends partly on whether it is true, which depends on later events. Something similar is often true of intentions. People frequently act intending in part to help achieve some goal in the future. Whether this intention adds to their perfection now depends, given a success view, on whether the goal is achieved, which again is settled later.

Given these facts, success versions of perfectionism can require people to promote goods in the past. Imagine that up to now a person has pursued a certain plan of life, say, that of living life *x*. This fact gives him a (defeasible) reason to continue with this plan of life. In the past he will have acted intending in part to live life *x*. But whether this contributes to his past perfection depends on whether he does live life *x*, which he can partly determine now.

The same can be true of particular intentions. Imagine that a poet writes thirteen lines of what she intends to be a fourteen-line sonnet expressing a certain thought. If her thirteen lines are good, the intentions behind them have some quality and her activity of writing them already has some worth. But it will have more worth if its final aim, a complete poem, is accomplished. This gives the poet a special reason to finish her poem. Not only is this a worthwhile activity in the present, but it also prevents her past activity from being in an important respect wasted.

These points about time increase the vulnerability of perfection to bad luck. In a success perfectionism, a person's excellence may be reduced by bad luck not just now but in the future, including at times after his death. If he devotes himself to a political cause that after his death succeeds, his activity may have considerable value. If the cause fails, however, his excellence is less. This vulnerability can create, if not self-regarding duties, then other-regarding ones. Imagine that our poet writes a series of poems, partly for their own sake but partly to share experiences with those who will read them when they are published. She dies unexpectedly, however, before publication is arranged. Here her friends have a reason, derived from her good, to publish her poems. Publication will ensure that her writing activity, which already has some value, has the full value that comes from having all its major aims accomplished.

Finally, the success views make distinctive claims about death. On the standard view, death is bad because it diminishes value in the future: It prevents people from being in the good states they would be in had they remained alive. Perfectionism accepts this view, although with its own account of the relevant good states. In its success versions, however, it goes further: It says that death is sometimes also bad because it diminishes value in the past. This is especially true when a young person dies prematurely, before achieving his major life aims. At the time of dying, this person will have ongoing projects. He will be taking the first steps towards goals whose realization lies in the future. In a complete life these preparatory activities would have value as parts of a long-term successful plan, but if death comes early their worth is diminished. The person's most general aims remain unfulfilled, and all their quality is lost. We commonly think that premature death, especially of someone with great prospects, is a tragic waste. Success versions of perfectionism explain why. According to them, an early death deprives not only the future but also the past of value.[17]

8.4.4

Because both views about relations have attractions, it is no surprise to find both in the tradition. Kant characterizes perfection non-relationally, saying a good will "is not good because of what it effects or accomplishes or because of its adequacy to achieve some proposed end," but has "its full worth in itself."[18] Plato and Aristotle, by contrast, value relations. For theoretical perfection they require knowledge and thus true belief, and at least Aristotle values success in action. This is evident both from his account of practical wisdom, which requires not just taking the right means to one's mark but "hitting" it,[19] and from his claim that a person's *eudaimonia* can be affected by events after her death.[20] To be consistent with the view that *eudaimonia* is an activity, this claim must value the successful pursuit of goals beyond one's death. Finally, relations are explicitly valued by many Idealists. Responding to Sidgwick's denial that "objective relations" to things outside the mind are intrinsically good, Bradley, Green, McTaggart, and Rashdall all insist that such relations are essential to human perfection.[21]

8.5 The Best Units?

Combining the presence and absence of two conditions we have four views of number, from attempt at one extreme to deserved success at the other. I have tried to show that each view has distinctive implications and also distinctive attractions. Is there any deeper reason for preferring one view to the others?

There might be such a reason if sophisticated psychology, which shows rationality to be essential to humans, clearly used one principle of charity (4.2.3). If rational interpretations had to maximize just the number of a person's consistent beliefs, this would favour attempt; if they had to maximize just true beliefs, this would favour success; and so on. As we saw, however, it is hard to argue that one principle of charity is superior to all others (4.2.3). Belief, justification, and truth are all relevant to psychological explanation and could be acceptable maximands of interpretation. Until one is shown to be a better maximand, we have some conceptual play room (2.3.3) and the need for a moral choice.

I see two main options here. If all four views have attractions, we may try to capture them all by adopting a mixed view, for example, one that treats every sincere belief as 0.2 of a perfectionist state, every justified or true belief as 0.6 of a perfectionist state, and every item of knowledge as a full state. (In each case this fraction is multiplied by a standard measure of quality.) This view gives some weight to each of the conditions imposed by the simple views, but never exclusive weight. It finds some value in the quixotic pursuit of unattainable ends, but more value in the successful achievement of ends, even somewhat lower-quality ends. It grants Aristotle some theoretical perfection for having justified beliefs about biology, but still sees a loss in the beliefs' falsity. It gives some importance to luck, but not overwhelming importance.

The other option is to make an intuitive choice among the simple views. (Perhaps there is an intuitive preference for a theory with a simple set of number-conditions.) Here my preference would be the most restrictive view, deserved success. This view supports pragmatism and the need for effective means, both of which are, in my view, endorsed by our most serious judgements of value. It also makes strong claims about consistency and condemns the experience machine. Deserved success makes perfection less self sufficient than it might be, but this may be acceptable if we distinguish carefully between judgements about "good" and "right" (5.1.2). While regretting the falsity of Aristotle's biological beliefs, for example, we can still greatly admire the choices he made in forming them.

Partly for simplicity's sake, I will follow the second option in what follows. I will assume that our view of number is deserved success, so our account of quality applies only to states that meet its twin conditions. But those drawn to a different view—a mixed view or a rival simple view—need not feel excluded. One merit of distinguishing number and quality is that we can discuss each dimension on its own, apart from debates about the other. Someone with a different view of number—and several are plausible—can substitute it for deserved success in what follows, and take as Aristotelian the slightly different moral judgements it supports.

9

Unity and Complexity

Perhaps the most important part of Aristotelian perfectionism is its account of quality, which identifies the kinds of knowledge that are best and the ends that are most worth achieving. What does Aristotelian quality consist in? What determines the value of a state once it has passed our tests of number?

9.1 Generality: Extent and Dominance

9.1.1

An acceptable account of quality must respect the concept of rationality and any limitations it imposes (2.3.3). Now, whatever the details of rational explanation—and not all proved to be determinate (4.2)—its general approach is *formal.* It says not that humans necessarily have beliefs about some specific subject matter or necessarily pursue some end, but that they have mental states with a certain formal character. To reflect this fact, an Aristotelian account of quality must likewise be formal, evaluating beliefs and intentions by their possession of some formal property. This formal approach is not universal in the tradition. Many perfectionists, including many who claim to value rationality, use *material* tests of quality. They say the best knowledge is of certain exalted beings, such as God or the heavenly bodies, and that the best acts pursue certain ends. In particular, moralistic perfectionists say that the best acts aim at the good or at obedience to certain rules. Earlier I argued that these last claims are inadmissible in narrow perfectionism: In any sense in which it is plausible that humans are essentially rational, it is not plausible that rational acts must be moral (2.3.2). Our account of quality must therefore reject moralism and, indeed, any material view of rationality. It must measure the worth of beliefs and ends using only formal criteria, and none tied to their specific content.

In developing this account, we need not retreat to a minimalist, or Humean, view of rationality. According to this view, rationality is entirely a matter of factors captured by the dimension of number: A belief counts as rational when it is sufficiently grounded in evidence, and an act is rational when it is well suited to its end, whatever that end may be. This Humean view is hardly sufficient for a plausible Aristotelian perfectionism. No one's ideal of the perfect life is exhausted by its containing justified beliefs and prudent ends, no matter what their content. Nor is the Humean view the only one possible. Merely forming a belief or intention requires rationality, and it can require more or less rationality. Humans are distinctively rational, with capacities beyond animals, because they can have and use mental states of a sophisticated kind (4.2). We can exploit this fact in characterizing quality. We can say that humans exercise rationality more, and are therefore more rational,[1] when their intentional states are more sophisticated. Because sophistication is a formal concept, this proposal respects the concept of rationality, but it gives us something richer than the Humean view. Although the Humean conditions are still required on the dimension of number, we can prefer some states that meet these conditions because they have more of our favoured formal properties.

This is the possibility I will explore in the next two chapters: a formal account of quality, based on formal properties of beliefs and intentions. In so doing I will define a distinctive *Aristotelian view of rationality,* one that falls between two more familiar extremes. It is not as rich as material views that demand pursuit of the good or as meagre as the Humean view. As I will show, the view supports attractive judgements of excellence; but it also captures a common understanding of rationality and deserves attention outside perfectionist theory.

9.1.2

What determines the intrinsic value of an item of knowledge? Again it is helpful to examine our intuitive judgements. In introducing the dimension of quality, I said it is better to know a fundamental law of the universe than the number of redheads in Beiseker, Alberta (8.1.1). If asked why, we will surely point to the first truth's greater *generality*. Whereas the claim about Beiseker is isolated and particular, a scientific law is highly general—in two different senses. In one sense, a scientific law is general because the state of affairs it describes is greatly extended. It includes all the objects at all the times in history, whereas a truth about Beiseker involves just a few people now. A law can also be general because of its place in an explanatory hierarchy. Someone who knows it can use it to explain many other truths, which then become subordinate to it in her theory of the world. It gains an explanatory generality not shared by the Beiseker truth. At issue in this second contrast is actual, not potential, explanatory importance. How much a belief *can* explain may depend on its generality in the first sense, that is, on its extent; how much it *does* explain depends on how it is used and, in particular, on how many other beliefs a person derives from it. For this reason the two kinds of generality can diverge. Sometimes an extended belief, for lack of the right accompanying beliefs, cannot be used to explain anything; at other times highly localized knowledge, say, about a historically important event, is rich in explanatory consequences.

The generality tests are also attractive for practical perfection. Earlier I contrasted a person who by elaborate planning achieves major political reforms with someone who merely ties a shoelace (8.1.1). If we ask why the first person's achievement is better, we can again say that his end is more general, first, because it is more extended. It includes everyone in his society for the foreseeable future, whereas the lace-tier's involves just one person and a shoelace now. The reformer's end also has more hierarchical importance. To achieve it, he must devote a large portion of his life to it, pursuing many other ends as subordinate means to it.

Whereas he develops a complex goal-structure with reform at the top and many others below it, the lace-tier's end is accomplished in a few bodily movements. As with theoretical perfection, the two kinds of generality can diverge. Sometimes a powerful politician can make far-reaching changes simply by issuing a decree, but placing a specific person atop Everest requires tremendous planning. Although it is barely extended, this last goal has many others subordinate to it in a means-end hierarchy and in the second sense is highly general.

9.1.3

Let us label these senses *extent* and *hierarchical dominance*. An intentional state is extended if its content stretches across times and objects, including persons; it is dominant if it has many others subordinate to it in a rational hierarchy, either because it is used to explain them (beliefs), or because they are means to it (intentions). Because these notions are formal, an account of quality built upon them is consistent with narrow perfectionism, and it may be more than just consistent. In chapter 4 I argued that humans' essential rationality includes a capacity for mental states that are extended and arranged in complex hierarchies (4.2). This claim passed the intuitive test, was implicit in sophisticated psychology, and would be even more compelling if charity required—as it may do (4.2.3)—that we maximize generality in the agents we interpret. The generality tests fit this claim and may, more strongly, be entailed by it. If humans necessarily have general beliefs and ends, it may follow that they are more perfect when their states are more general. The issue, however, is difficult. If a property does not figure at all in rational explanation, it clearly cannot determine Aristotelian quality. But it is harder to argue that what does figure in rational explanation must determine quality, and I will not so argue. I will not claim that the generality account is uniquely admissible in narrow perfectionism, but will develop it as one of perhaps several possible standards of quality. Assured that extent and dominance are consistent with narrow perfectionism, I will support them only by pointing out their intuitively attractive consequences.

9.2 Generality: Elaborations

To apply the generality account, we must clarify its component measures. Of these, extent is the simpler measure. Some mental contents have fuzzy boundaries, and there are issues about the relative weights to be given extent across times, objects, and persons. So, as elsewhere in perfectionism, we cannot expect precise measures of extent.[2] Dominance, however, raises more complex issues.

The measure of dominance values intentional states that have many other states subordinate to them, and therefore indirectly values the hierarchies that embody this subordination. Consider a particular structure of beliefs or intentions (Figure 9.1). It contains more dominance than seven unconnected states (Figure 9.2) and therefore is better. But exactly which hierarchies have value, and how do states contribute to this value?

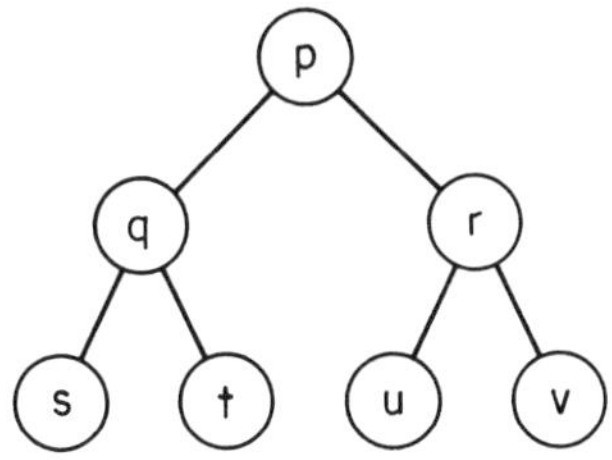

Figure 9.1

9.2.1

Some initial restrictions come from the dimension of number. Having accepted the deserved success view (8.5), we count only beliefs that are justified and true and only ends that are non-luckily accomplished. By extension, we should let only states that pass the same tests contribute to dominance. In calculating the dominance of an item of knowledge, for example, we should count only beliefs subordinate to it that themselves count as knowledge. In Figure 9.1, a person may know truth *p* and believe that *p* explains *q*, where *q* is false. This fact should not increase *p*'s dominance: What does not count for number should not be relevant to quality. Nor should an end *r* be better if, before accomplishing it by means *v*, a person tries *u* but cannot bring about *u*. What is intended but not accomplished as a means should not be valued as a means, that is, should not make a final end more dominant.

The same restrictions should govern the relations between states. Someone may know truths *p* and *q*, that is, have genuine knowledge of both, but believe falsely that *p* explains *q*. This fact should not increase *p*'s quality, nor should an end *r* be better if initial means *u*, although successfully accomplished, does not help to bring about *r*. Hit-and-miss procedures, where one tries many means that lead nowhere, should not have significant value.[3]

9.2.2

To avoid infinite measures, we had the dimension of number count not individual states but states *of different kinds* (8.1.4). We need the same device within dominance, for the same reason. Used in the right way, however, it has further attractive consequences. Let us assume that a state's dominance equals the number of states of different kinds that are subordinate to it, and that some of these kinds include others, as sports truths include hockey truths and golf truths. Then hierarchies are better the more varied their constituents or the more diverse the states they combine.[4] The greatest dominance springs from the richest variety in subordinate beliefs and ends.

This claim is intuitively appealing. Many of us think the best knowledge ex-

Figure 9.2

plains, not just many truths, but truths in different areas. It draws surprising connections between phenomena, uniting facts that at first seemed unrelated. In the same way, the best practical endeavours have varied subplans, requiring actions of different types. This variety is often found in politics, where a reformer may have to master economic theory, negotiate agreements, raise funds, and deliver speeches, all as means to a final political goal. By contrast, variety is absent from repetitive manual labour. A worker on an assembly line makes many movements as the means to earning his day's pay, but all are essentially the same. All are movings of this lever or pushings of that knob. If dominance counts kinds of subordinate state, it finds this labour low in value. The many individual ends in the worker's hierarchy are of few different kinds, making for limited dominance.

9.2.3

A third issue concerns time. In calculating a state's dominance at a time, we can count only those states subordinate to it at that time or all the states subordinate to it in a person's life as a whole. On the first, or *present-subordination,* view, it is irrelevant to a belief's quality now that she did earlier or will later use it to explain truths she does not now know. Given *life subordination*, however, subordinate states at other times count as much as those now present.

Of these two, I prefer the life-subordination view. By making it impossible to judge a state at one time without knowing facts about related states at other times, this view strengthens perfectionism's emphasis on the complete life as a morally significant unit (2.2.2, 5.2.2, 6.1–6.3, 7.4.2). At the same time, it gives perfectionism more dramatic consequences. Imagine that a person pursues the ends in Figure 9.1 sequentially, achieving *s, t, u,* and then *v,* producing *q, r,* and finally *p.* Assume also that each state earns one unit of dominance for being subordinate to itself, plus an extra unit for every other state subordinate to it in a hierarchy.[5] At each time, the person pursues one end in each row in Figure 9.1. Given present subordination, the bottom-row end has a dominance value of 1, the middle-row end a value of 2 (itself plus one bottom-row end), and the top-row end a value of 3 (itself plus one end from each of the middle and bottom rows). This calculation gives the person 1 + 2 + 3 = 6 units of dominance per moment, which contrasts favourably with the 1 unit per moment achieved by someone with seven unconnected ends (Figure 9.2). But the difference is even greater given life subordination. Then the bottom-row end has a dominance value of 1, the middle-row end a value of 3 (itself plus two bottom-row ends), and the top-row end a value of 7 (itself plus all the other ends in the hierarchy). The person achieves not just 6 but 1 + 3 + 7 = 11 times more perfection than someone with unconnected ends. Something similar occurs in the intermediate case where a person achieves six separate goals as means to an overriding seventh (Figure 9.3). Here present subordination finds 1 + 2 = 3 units of dominance at each time, whereas life subordination finds 1 + 7 = 8. This total is less than the eleven units in a more structured activity (Figure 9.1), but again the life view magnifies a value difference. In what follows I will adopt this view, equating a state's dominance with the number of states of different kinds subordinate to it at all the times in a life, and accepting the strong conclusions it implies (9.4.1, 9.4.3, 10.4.4).

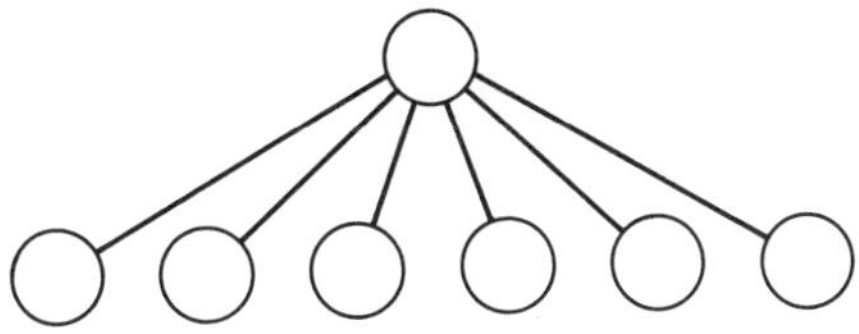

Figure 9.3

9.2.4

Given life subordination, a state at one time can affect the value of a state at a different time. To do so, however, it must be connected at its own time to the other state: an end must be intended *when it is pursued* as a means to one that is more general, and a belief must be derived *when present* from one that is more explanatory. This condition excludes a proposal made by Gilbert Harman: that it is rational to adopt ends now that will organize unconnected ones from the past or turn them to a new general purpose. Consider someone who studied physics at university because he liked it or because physics was a required subject. According to Harman, this person has a special reason to become a physics teacher. Doing so will make sense of his past decisions by revealing them as means to a goal he adopts only now. By teaching, he "gives significance to his earlier study and to that extent unifies his life."[6] I do not believe we should accept this idea. If the student's courses were not intended when he took them as means to teaching, they should not contribute, even given life subordination, to his present goals' dominance. To be relevant to the value of a hierarchy, lower elements must be viewed, when present, as belonging to that hierarchy.

9.3 Top-to-Bottom Knowledge

We can now apply the generality account and see what specific states it values. As we proceed, we will often find extent and dominance favouring similar goods. They do so for different reasons, however, and each has its own distinctive implications.

9.3.1

A first application is to theory. Here a person can achieve both kinds of generality by grasping a scientific theory all the way from general principles down to particular explanations. Greatly extended knowledge is good, as is explanatory integration. But a person has both if she knows a fundamental law and has done the work needed to apply it to particulars. Then both the law's scope and its explanatory importance are reflected in her mind. She knows a truth that governs many objects, but has not rested there. She has used it to explain many facts about these objects, realizing an ideal of *top-to-bottom knowledge*.

This ideal unites two distinct styles of inquiry, one concerned with abstract, unifying principles and the other with detailed, particular facts.[7] In our time, the ideal may not be fully realizable by any one person. Our most fundamental explanatory principles are so abstract and difficult to grasp that learning them leaves little time for acquiring particular knowledge. At the same time, the wealth of information to be had about, say, plant life in a particular forest is so great that it precludes much attention to foundational laws. Scientists today must therefore choose between aspects of the ideal—between having the most extended knowledge and explaining the greatest number of facts. No one choice here is always correct. Some do best by speculating about fields and elementary particles, others by exploring the details of biology. But if both achieve some theoretical perfection, it is because both realize aspects of a single ideal defined by extent and dominance.

9.3.2

If knowing a unified theory is good, it is even better when one's knowledge is precise. Someone can know vaguely that the gravitational constant is between 8.8 and 10.8 m/sec^2, or know exactly that it is 9.8 m/sec^2. In the second case he knows more truths and so scores higher on the dimension of number. He knows not only that the constant is between 8.8 and 10.8 m/sec^2, but also that it is between 8.9 and 10.7, 9.0 and 10.6, and so on. For the same reason, he can explain more truths. Knowing the precise value of the constant, he can explain why an object released towards the earth two seconds ago has a velocity, not only between 17.6 and 21.6 m/sec but also between 17.7 and 21.5, 17.8 and 21.4, and so on.[8] The second point is the crucial one. There are many precise truths, for example, about redheads in Beiseker, that are not especially worth knowing. When a general law is precise, however, and the facts derived from it are also precise, dominance increases throughout a hierarchy. We have seen how dramatically hierarchy changes can magnify excellence—eleven times in a simple example (9.2.3). We can expect similar effects when the changes involve more precise—and thus more—truths at every level of a structure. For dominance the best knowledge is not only systematically organized but also composed at every point of precisely formulated truths.

If the highest ideal is unified scientific knowledge, there are also more modest theoretical goods. Consider an artisan with a subtle understanding of the quirks and capacities of her material, or a parent who understands exactly how his child feels and why. Each has some moderately extended principles, with some particular insights below. Each can make precise discriminations and explain why something is this way and not that. Perfectionists writing about knowledge often concentrate on its most rarefied forms, such as theology (if that is indeed knowledge), philosophy, and physics, but outside these abstract domains are many bodies of knowledge with some generality and some intrinsic worth. They do not differ in kind from scientific understanding, but embody the same virtues to a lower degree. We must not think that if perfectionism values knowledge it will care only about the activities of an elite crew of researchers. If its aggregative principles are distributively neutral (6.2.2), it will want to ensure some knowledge for all and may prefer extending modest understanding to the many over improving the exotic insights of the few.

9.4 The Unified Life

The generality account has several practical implications.

9.4.1

For one, the account values any life that is organized around a single end or in which large parts have a single end. This life is unified and therefore more rational than one lived only day to day. It expresses a coherent personality as disconnected experiences cannot. Of course, some ends, such as knowledge or athletic excellence, are better than others, and lives organized around them are for that reason more perfect. But the present claim abstracts from this fact to say that any unified life, regardless of its content, derives some value from its unity: Its very organization gives it a kind of perfection. In the real world, of course, no lives are fully unified, and the issue is only how closely we approximate to an ideal. Without organizing a whole life, we can organize a year or month, or we can have one goal for our career, personal life, or leisure. For Aristotelian perfectionism, living rationally means living by a plan, and the more we do so, the greater our practical perfection.

The chief value here is dominance. A single life-defining end has subordinate to it all the others a person achieves in her life (9.2.3) and is therefore enormously general. This is true not only when the subordinate ends come in their own structures but also when the person lives always for the present, believing this the best life. Then she instantiates the structure shown in Figure 9.3, with many separate ends subordinate to a single one that overrides. This life is less perfect than a life of greater complexity (Figure 9.1), but its unifying purpose makes it better than one lived day to day without a thought for why (Figure 9.2).

At the same time, a unified life involves considerable extent. Often a person's organizing goal is to live a certain kind of life; even when it is not, such a goal comes near the top of her hierarchy. But this kind of goal is extended through time, involving states of the person at all the times she lives. To structure one's life is also, necessarily, to seek a goal stretching from one's past into the future.

These points strengthen a claim made in the previous chapter. Given a success view of number, people always have some reason to continue with a plan of life pursued in the past (8.4.3). The strength of this reason depends on the quality of the end that defines the plan, and this quality, we now see, is considerable. As doubly general, it has, when achieved, considerable worth. The reason to persist with the plan is defeasible: A unified life is just one good among others and can be outweighed by greater goods. Given generality measures, however, it does embody substantial quality.

These points also vindicate an assumption made at the start of chapter 6. A unified life clearly cannot be lived at one moment, but, given Aristotelian measures, its value can be seen as composed of perfections achieved at particular moments. At each time, a person living a unified life intends in part to do so, but how this intending affects her perfection at that time depends in several ways on facts about other times. It depends on whether she succeeds in living the life she intends (8.4) and on how many other ends are means to this life at other times (9.2.3). The unity of her life is achieved through time, but its value appears in measures at particular times.

9.4.2

Even a highly unified life need not be monotonous or lack variety. First, nothing in the concept of an extended goal prevents the goal from involving change. Someone who plans to switch his careers every five years still has an end that stretches through time, even though its elements are diverse. Second, the measure of dominance positively values variety. The Aristotelian ideal is not just unity, which is possible in a life spent idling or sleeping. It is a unity of substantial elements, namely, those that comprise a subordinate hierarchy. And if these elements are counted in kinds that can include each other (9.2.2), the hierarchy has more value the greater its variety. In its best formulation, the Aristotelian ideal is not unity but unity-in-diversity, or the bringing of many contrasting elements into one life structure.[9] Far from excluding richness, this ideal requires it for a life's highest worth.

A life's unifying end also need not be monolithic. Imagine that someone enjoys philosophy and hockey but knows he cannot pursue both at once. One way he can organize his life is to devote himself entirely to one activity, say, philosophy, and abandon the other. Another is to conceive a life containing both activities, in some more or less definite proportion, and aim at that. The second approach is no less valuable. It still involves a single life-organizing end and particular choices to achieve it; it still stamps one character on a series of acts, now that of a composite ideal. In some respects the second approach is preferable: It allows more balance between perfections (7.2) and a more varied hierarchy. The important point, however, is that it is allowed. For Aristotelian perfectionism, the important difference is not between those who do and do not choose one activity above all others. It is between those who, seeing their alternatives, do and do not form a settled plan for combining activities within their life.

9.4.3

Closely related to a unified life is a further aspect of prudence (8.2.2). For many philosophers, prudence involves not just effective means to one's present ends but concern within those ends for one's future good. It involves an impartial concern for all the parts of one's life, regardless of their temporal location.[10] This concern is integral to a unified life. Someone who aims to live a certain complete life will care equally about all his days as constituents of it. But impartial concern is also possible outside a unified life. Consider a person whose ends are in constant flux, who acts always on whatever whim strikes him. If he recognizes his inconstancy, he can act to ensure that he will always be able in the future to pursue successfully the goals he then adopts. If so, he exercises future-directed prudence, even though his life has littlc unity.

If unity is good, there is intrinsic disvalue in whatever detracts from it. One such failing is weakness of will. Its irrationality may be captured partly on the dimension of number, if weak agents have conflicting intentions (8.3.3). Given the generality account, however, it has a further, more important defect. The end a weak agent acts against—for example, self-interest or living a good life—is usually extended, concerning large parts of the future. It is usually also an end she takes other means to at other times, so there is already a large hierarchy of ends below it. This means the end would be highly general if it were acted on now, so her weakness involves a loss of quality now. At the same time, her weakness diminishes the end's quality at other times. If she did not act weakly, her present conduct would fit into an ongoing project and, given life subordination (9.2.3), would increase the dominance of its defining ends at all the times they are sought. It would make the project's whole hierarchy larger and its perfection greater. By succumbing to weakness, the person abandons unified activity for a particular momentary goal and so severs the ties between her acts. If weakness of will is intrinsically bad, this disunity is surely the main reason. More telling than any loss on number is the fact that weak acts fall outside a life's main projects and diminish its overall coherence (see further 11.1.2).

9.5 Complex, Difficult Activities

The next claims concern, not the overall structure of a life, but its choice of particular activities. A life must, however, contain activities. With intentions as our general practical category, particular practical goods must involve the active pursuit of goals. But which particular goals are best?

9.5.1

The generality account values activities that, as far as possible, are complex, intricate, and challenging. These activities stretch our capacities, demanding more rationality than ones that are simple. Chess at the grand master level is difficult, as is mountain-climbing. Athletes such as Wayne Gretzky solve sophisticated tactical problems during their games, as do scientific researchers and politicians. In their different ways, these agents all have challenging projects and achieve more than they could by doing something easy. Rawls too values difficult activities, citing in particular the superiority of chess to checkers, but his reason is that, given his Aristotelian Principle, these activities are more enjoyable.[11] We have rejected any strong Aristotelian Principle (3.1.4) and must value difficulty directly. There are three grounds for doing so.

First, difficult activities usually involve a more complex subordination of ends and therefore involve more dominance. To achieve checkmate, a grand master adopts a series of substrategies aimed at developing his pawn structure, creating attacking positions, and so on. Each substrategy has further goals subordinate to it, and the result is a more ramified structure than anything needed in checkers. The same is true of other activities. Mountain-climbing requires many steps before feet reach a peak, and because each aspect of a trek—equipment, route selection, conditioning—has component parts, again complex relations among ends are involved. A painter uses all his individual strokes as parts of a single composition; a criminal lawyer, following the twists and turns of a trial, intends her many ploys to create a combined effect on a jury.

In one sense, complex activities are difficult because there are many ways they can go wrong: The more elements they contain, the more chances for failure. Complex activities also make greater demands on rationality. Sophisticated problem-solving means going through stages, doing one thing now so that another can be done later. It means grasping a plan in its entirety and monitoring progress through it as one accomplishes its parts. Both these tasks are harder when the plan is more complex. More elaborate structures are harder to formulate and hold in the mind, especially when they are more varied (9.2.2). Even good checkers players consider possibilities that are of essentially the same kind and can be run through mechanically. But chess players have pieces with different powers and must bring to bear their different understandings of attack and defence, openings and endgames, plus whatever they know of their opponent's psychology and past patterns of play. Balancing a range of considerations, they must manipulate a varied hierarchy.

Second, the complexity of difficult activities is often accompanied by greater (non-temporal) extent. Many checkers players move whichever piece has the best individual prospects, with little thought for the effects elsewhere. Serious chess players, however, consider the whole board. The states they bring about are not localized but involve all the pieces in play, understood as forming an extended matrix of forces and possibilities. Grasping these extended states is difficult, and chess players learn to do so only with practice. Only after repeated plays can they see positions in their entirety and extract in a unified way the information they contain. Nor is this phenomenon unique to chess. Compare a musician's ability to hear a melody as one unit and reproduce it correctly on first hearing, or a logician's ability to see the structure of a whole proof. Think, too, how Henry Moore sees a sculpture:[12]

> He thinks of it, whatever its size, as if he were holding it completely enclosed in the hollow of his hand; he mentally visualizes a complex form *from all round itself;* he knows while he looks at one side what the other side is like; he identifies himself with its centre of gravity, its mass, its weight; he realizes its volume, as the space that the shape displaces in the air.

The great ability of athletes like Gretzky is to see all the game around them, know where other players will be next, and act on that knowledge. This ability is reflected in their tremendous recall of patterns from completed games.[13] In all these cases, there is an ability, partly innate and partly trained, to stretch the mind around extended states and compare them in the light of global properties. This stretching is a high exercise of rationality and immensely difficult.

Finally, difficult activities usually involve precision. They require not some subordinate means or other, but precise steps taken in precise ways. The value of this precision is the same as within theoretical perfection (9.3.2). When I shoot the puck in hockey, my intention is usually to put it somewhere near the net. When Gretzky shoots, he intends not only what I intend but also to pick the spot the goaltender has left open, which is in the upper right corner, and a few inches across. He has all my intentions plus many more, namely, those that specify his precise goals. Because of this, he accomplishes more goals in a game than I do and, more importantly, has more goals subordinate to his main one of winning. The same holds in other activities. Skilled chess players advance their bishops not just at some moment but right when an opening exists, and the challenge of mountain-climbing is precisely that so few routes lead up Everest. Easier activities do not require this precision. In them we can make do with vague subordinate ends—hitting some part of the barn door—and need not add the extra ends that make for precision. By giving us leeway with our means, these activities require fewer ends and a less elaborate hierarchy.

These three aspects of difficulty—complexity, extent, and precision—are independent of each other, and an activity can have some of them but not others. Painting by numbers and building model airplanes involve precision but lack extended states or complexity; politics can be extended but not precise, and so on. The very best activities score high on all three dimensions: Think of chess, military strategy, and novel-writing. But again it is wrong to focus only on the highest goods and ignore ones lower down the scale of value. More modest practical achievements are possible in gardening, child rearing, and carpentry; they are also possible in hosting a party. In each of these activities, particular cases differ: Just as one can play better or worse chess, so one can approach a garden or the education of one's children with more or less care. But that some value is possible in everyday pursuits is important. Otherwise, only a few people would have talents worth developing; as it is, there are serious practical goods to be sought for all.

9.5.2

Chess players usually manipulate their ends consciously, so they are aware of what they intend. But conscious ends are not necessary for practically perfect action. The dimension of number counts dispositional and therefore unconscious intentions: If these are present in the right way, with the right formal properties, they can embody extent and dominance.

Unconscious intentions are a feature of traditional craft activity. A traditional wheelwright, we are told, could reject likely-looking wood as unfit for use without being able to name its exact flaw, and when constructing a wheel could maintain the right proportion between spokes and felloes without any equipment or formal rules. His feel for the proportion was not conscious but "in his bones."[14] The wheelwright had to use some general principles; how else could he achieve consistent results? But his principles were not present to his consciousness, nor were the hierarchies they helped to define. In the same way, many athletes make split-second decisions during games but cannot explain afterwards what they did. If their acts are intentional—which, watching, one cannot doubt—their value is undiminished. Unconscious ends are still ends; they can still produce behaviour in sophisticated ways.

The unconscious rationality of craftsworkers and athletes has some limitations. Although some ends can be handled without awareness, this is harder to imagine for the complex strategies of chess and novel-writing. And craftsworkers who are entirely unreflective may fail to arrive at some valuable knowledge. If they never reflect on their principles, they may never discover the deepest explanations of why their choices succeed. But neither point touches the practical perfection they do achieve, and the intuitive approach has compensating advantages. Unconscious deliberation is faster than that involving reflection and better suited to activities where the premium is on speed. It is also easier to insulate from the disrupting effects of ego involvement, wishful thinking, and nerves. For this reason, some writers claim that mastery in a craft or sport requires the elimination of consciousness, so all is done unreflectively.[15] For our purposes, whether this claim is true is not so important. What matters is that unconscious hierarchies can have the same value as ones accompanied by awareness.

9.5.3

Like a unified life, difficult activities are better when they aim at something independently good. Their having such an aim is a merit of scientific research and child rearing: As challenging as other activities, they also produce new knowledge or a good human life. But an independently good product is not needed for practically perfect activity, which is possible without it. This is clearest in games. A chess player's goal of checkmate is intrinsically trivial, as is that of a mountain-climber or hockey player. There is no value in standing atop Everest if one has been deposited there by helicopter and nothing intrinsically worthy about a puck's entering a net. In games, people adopt inherently valueless ends in order to exercise skill and ingenuity in achieving them. They rule out the simplest means to those ends—moving the other player's pieces or flying up Everest—and take only the difficult means the game's rules allow.[16] The point of these restrictions is captured perfectly by formal measures of quality. If the value of an activity is partly independent of its content, a game requiring us to pursue general ends can be excellent even if its final aim is trivial. In fact, games are a paradigmatic illustration of Aristotelian values. In them it is best to achieve a goal, so that one's activity has a product; but the value of this product depends entirely on features of the process that produced it.

9.5.4

For many, the value of playing a game depends partly on the attitude of the participant. If someone plays chess because he loves it or to exercise his intellect, his activity has considerable worth. If he plays only for money or to win, however, his activity does not; even with the same skilful tactics, his play counts for less. As we have developed it to this point, the generality account does not capture this view. It says that someone who plays chess only for money lacks proper values and a proper appreciation of his talents, but not that he achieves less perfection. An emendation to the account will yield this result.

A practical hierarchy can contain subordinate ends that are intended only instrumentally, that is, as means to something else, but it can also contain ends that are chosen partly for themselves. Although contributing to ends beyond them, these second ends might also be chosen alone. They are intended for two reasons: as means to a further goal and for their own sakes.

These doubly intended ends are special, and hierarchies containing them are more complex: They contain not one but many ends willed for their own sakes. This makes it natural to give such ends greater perfectionist weight. One way of doing so is to hold that a subordinate end intended partly for itself generates a separate hierarchy, with a separate calculation of dominance. Imagine that a person achieves *r* as a means to *q* and *q* as a means to *p*. If he wills *q* and *r* purely instrumentally, his dominance values are $1 + 2 + 3 = 6$ at each moment (9.2.3). If he chooses *q* in part for itself, however, he gets the six units in his *p-q-r* hierarchy, plus three more in a *q-r* hierarchy. If he also chooses *r* for itself, he gets an extra unit for that. There is double counting of subordinate structures whose highest elements are doubly intended, or double dominance for means that are also willed as ends.

This emendation has the desired implication about games. Someone who plays chess only for money views his moves just as means to a further end, so in his hierarchy they count for less. Someone who loves the game for itself, however, intends the same moves intrinsically. For him the skilful deployment of his pawns and the timely castle count not just as means to winning or playing a good game, but as themselves exhibiting chess ability. In his hierarchy they generate separate calculations of dominance, yielding higher practical perfection.

The value of double intending must, of course, be weighed against other perfectionist goods. Given the value of a unified life, it seems best to play a game not just for love, but for both love and money. Then an intrinsically valued activity is also connected to larger concerns. But what if the two attitudes are incompatible? What if any concern for money will drive out double intending? If so, the highest achievable good may be to play the game only for itself.

Another issue concerns the relation of double intending to a player's technical skill or achievement. May not a concern for winning—even an exclusive concern for winning—be necessary for the highest levels of performance? If so, we face a conflict between two ideals: of the amateur and of the professional, of loving the good for itself and of developing the highest skill. Neither ideal will uniformly outweigh the other; each embodies a value the other does not, and it will be a matter of seeing, in particular cases, which value is greater. In some cases, the conflict may be avoidable. In games with a premium on speed and aggression, such as hockey and sprinting, one's performance may be aided by visions of victory, but in other cases the opposite is true. Sports like figure skating and target shooting, which emphasize precision and consistency, can be spoiled by thoughts of an external goal. In them the intrinsic intending of the amateur is not a hindrance but actually needed for the highest technical achievement.

Double intending is equally possible outside games. Work, exercise, and personal relations are likewise better when their components are chosen partly for themselves, as are whole lives. Earlier, we considered a life with a composite goal mixing philosophy and hockey in a pre-intended proportion (9.4.2). A slightly different life contains the same blend of philosophy and hockey but uses the second only as a means to the first: Hockey is played only to replenish one's mental energy and to provide philosophical examples. This second life may seem more integrated than the first: Instead of two separate hierarchies converging on a composite ideal, it has all its ends in a single structure. But many of us would judge the first life as better, and our present idea explains why. When a person plays hockey only as a means, individual games and moves embody less dominance; when it is loved for itself, they count for more. Like particular activities, whole lives are best when their parts are willed intrinsically and not just as means.

9.5.5

Let us return to more general issues. We have seen that the practical generality characteristic of a unified life appears also in particular activities that are complex and difficult. In both there is the same sophisticated rationality, that is, the same achievement of extended, dominant ends. The generality account therefore unifies some diverse-looking goods: in commending an organized life, chess, and musical composition, it does not value heterogeneously but applies a single practical standard.

To make final judgements about lives, the account must be able to compare these goods, which is often not easy. It is often unclear whether the value in a unified life outweighs that in particular difficult activities or vice versa. Imagine that one life is organized around a single end but involves only simple pursuits, whereas another flits anarchically between different challenging projects. A realistic perfectionism may be unable to judge one of these lives as better than the other. What it can judge, and judge decisively, however, are the best and worst lives. Given our practical applications to now, the best life has a single organizing end that demands many different challenging activities that are also valued for themselves, whereas the worst is an unconnected series of passive experiences. In the first case, there are extended, structured goals both for one's whole life and at particular times; in the other, there are no such goals anywhere.

10

Politics, Co-operation, and Love

To this point our main claims about practical perfection have derived from the measure of dominance, with support from extent when that favours ends that stretch through time or involve many objects, such as all the pieces on a chessboard. But extent also values ends that involve many people. Achievements that encompass other humans take us beyond the boundaries of our lives and express a more expansive agency.

10.1 Political Action

10.1.1

These achievements belong, first, to political leaders or those responsible for large political effects. In favourable conditions, a political leader can knowingly change conditions in a whole society. By enacting new laws, she can alter some states of all her subjects: their wealth, their relationships with each other, the institutional structures they face. Her impact on her subjects would have no value if it were not intended, nor, absent romanticism (8.2.1), do mere intentions to have an impact suffice. But with sufficient power a political leader can formulate and realize ends that involve a whole populace. A similar extent is achieved by others such as factory managers, who also administer large numbers of people. Like a political leader although on a smaller scale, a manager conceives a plan involving all his workers and brings it to fruition. He too has an end that is too sophisticated for animals—one involving a whole group—and realizes it through action.

When we evaluate political leaders and managers, our main interest is, of course, instrumental. We want to know whether their decisions have been for the greatest good of society or of their factory. The measure of extent does not deny this interest. It is not a device for making all-things-considered judgements about people's acts; instead, it concerns just one intrinsic value in one person's life. And here it makes a distinctive claim: that any political or managerial act, whatever its final outcome, has some value in itself. By affecting many people, it embodies some generality.

Something like this claim is affirmed by many perfectionists. Seeking a rival to the contemplative life as a candidate for the best life, they cite not some life of practical activity but the political life in particular (7.3.4). What explains this choice?

Political activity is complex and difficult, but so are moneymaking, chess, and art. What distinguishes the former seems to be its greater extent. Political leaders paint on a larger canvas than ordinary citizens and affect more of the world. Their ends are more encompassing and to many perfectionists, it seems, more perfect.

10.1.2

The exact value of a political act depends on what other intentions it involves. Imagine that a political leader brings about a state in many people on a whim—because she feels like it and for no other reason. Then she intends one extended end but no others, and in particular no others in a hierarchy. This gives her activity less value than if she acted on general political principles, making all her subjects *F* so they could be *G,* which would permit their being *H,* which would realize a more general ideal. In this second case, she would have not one but several extended ends, and they would be arranged in a structure. Intuitively, a politician who acts on general principles exercises rationality more than one who merely follows caprice. Our generality measures explain why.

Even when it involves general principles, extent in politics differs from that in chess or music (9.5.1). A chess player who intends an extended matrix also intends many of its parts: He intends that his queen be here, his bishop there, and so on. Similarly, a composer intends all the individual notes in his composition. But the analogue is not usual in politics. A ruler does not usually know all the individuals her acts will affect and therefore cannot intend all the parts of what she achieves. This does not diminish her final ends' extent: Just as one can understand a scientific law and embody its generality without being able to identify all the objects it governs, so one can make many people *F* without knowing who all of them are. Still, there is a perfectionist loss. A ruler with general principles can have hierarchies among her extended ends, but she cannot have hierarchies within them. She cannot intend to affect many people by affecting a series of named individuals. This seems, however, to be just how our two measures relate. To achieve the most extended ends, as in politics, one must sacrifice the dominance that comes from intending their parts; to retain that dominance, one must seek the less expansive goals of chess or music.

10.1.3

Which political structure maximizes political perfection? Given maximax aggregations (6.2), the most desirable structure is one-person rule. With unlimited power, a monarch can achieve more politically than any citizen does when power is shared. But the best perfectionism does not employ maximax. It gives the same weight to perfection in all human lives (6.4) and therefore probably favours the system at the opposite pole from monarchy, namely, democracy. In a democracy no one citizen's political efficacy is as great as a single ruler's, but the sum or average of their efficacy is surely greater. If all citizens join in decision-making, all can have some impact on the larger world around them.

This argument does not provide a complete defence of democracy: That would have to consider the effects of democracy, not just on one perfection, but on them all. More specifically, it would have to show that democratically reached decisions are best at promoting the overall excellence of all citizens. Given a measure of extent, however, this justification has a distinctive subpart. It values democracy partly for letting all citizens play some role in political life.

It may be objected that in a representative democracy this role is so limited as to be valueless. There most citizens' political activity is confined to voting in a mass election every few years. Where is the perfection in a vote that normally has no effect whatever?

There can be indirect benefits from representative democracy. Universal suffrage encourages citizens to at least follow events on the national and international scene. They thereby gain some extended knowledge of human affairs, even if they cannot use it to achieve extended political ends. It may also improve their non-political ends by increasing their concern for others or their capacity to benefit them.[1] Finally, democracy can stimulate co-operation (10.2). In an election campaign, a party worker may canvass a poll as part of a co-ordinated effort involving his constituency organization and, beyond that, a nation-wide team. Between elections he may work with others in demonstrations and lobby groups to influence opinion or government officials. Given one-person rule, these co-operative activities would have little point, but democracy makes them common.

There could also be greater direct benefits from representative democracy. Writers who value political action have proposed reforms to increase the participation of citizens, such as neighbourhood assemblies to discuss national issues, computer polling of citizens in their homes, and greater use of national initiatives and referenda.[2] These devices would give citizens more political efficacy than does contemporary voting.

Still, even a reformed national democracy will offer most citizens only limited opportunities for effective political action. It is therefore no surprise to find, through much of the perfectionist tradition, a preference for small-scale, decentralized political units. This preference runs from Plato's and Aristotle's attachment to the Greek *polis,* through Green's wish to devolve important decisions onto town and county councils, to the enthusiasm of many contemporary socialists for democracy in the workplace or single industrial unit. And it seems plausible that political achievement will be greatest given some decentralized democracy, where people make effective contributions to decisions by small-scale authorities. This is, again, not a decisive argument for small-scale democracy. Considerations of efficiency, important for their bearing on other perfectionist goods, may require greater centralization. But if political action is one good, we may be willing to sacrifice some other goods for its sake. We may tolerate some instrumentally less good decisions if the process of reaching them was better. If decentralized democracy is almost as good as other structures on other dimensions, its intrinsic merits may make it the best structure overall.

10.2 Co-operation

10.2.1

People also pursue extended ends whenever they co-operate, working together to achieve a goal involving states of them all. Imagine that a group collaborates to understand a Platonic dialogue, win a volleyball tournament, or make their neighbourhood attractive. Even if many group members contribute mainly by realizing some state in themselves, their final ends are more general. If the group succeeds, it is true of each member that he has intentionally helped many understand Plato or many win or many live in an attractive neighbourhood. By co-ordinating his acts with others', he has increased his own generality.

For such co-operation to be valuable, each participant must help to achieve the extended end. At the minimum, he must bring about that part of the end that lodges in himself: that *he* understands Plato or that *his* house is attractive. If he does this, however, he gets credit for the end's entire extent. If four others come to understand Plato, his doing the same makes the difference between there being and not being five people who understand Plato. And if he intends that five understand Plato, he counts as achieving this whole extended state. So, of course, do they.[3] There are further gains in extent if all will their activity *as co-operative*. Then each intends partly to bring about a state where all intentionally bring about a state involving them all, and each gets further credit for doing so successfully.

The structure of co-operation differs from that of political rule: Instead of one person imposing states on many others, we have many people achieving an end together. But the value in the two cases is the same: that of extending one's ends to embrace other people.

10.2.2

Different styles of co-operation combine extent and dominance in different ways. At one extreme is co-operation "at a distance," where people knowingly pursue a common goal but do not interact directly to achieve it. In a neighbourhood some may see others keeping their properties attractive and do the same without consulting anyone. Similarly, researchers in one laboratory may see others studying a problem and join in without doing more than informing the first of their plans. Given the intention to contribute to a joint activity, co-operation at a distance still involves extended ends. The residents still intend that all keep their properties attractive, and the researchers that all work towards a final discovery. But these extended ends are isolated at the top of a hierarchy. After intending that all achieve a state, most contribute mainly by achieving just their own small part.

At the other extreme is "face-to-face" co-operation, as in a Platonic study group. Here continuing direct contact makes each person responsible for many extended states, as she affects the others in countless ways. She has not just a few extended ends, but many such ends in a hierarchy. We have here another case of conflicting values. Co-operation on the largest scale cannot be face to face or offer much dominance; the most intense collaborations cannot be very extended. Although neither kind of co-operation is alway better than the other, each expresses generality in a slightly different way.

Co-operation is a prominent element in many accounts of ideal human organization. Many accounts of "meaningful work" have as one component that workers understand their place in the larger structure of a firm and see how their labour advances its ends. This, we can say, is a matter of co-operation at a distance, of knowing and willing relations to one's co-workers. On a grander scale is the Idealist conception of an "organic society" and its associated ethic of "my station and its duties."[4] By playing a role in a structured society, Idealists claim, we connect ourselves to a larger whole and thereby increase our perfection. With at least the Idealists' conclusion we can agree, provided certain conditions are met. It is not enough for Aristotelian perfection that there be an organic society with complex relations among its parts. The members of the society must understand these relations and will their acts in part as sustaining them. Any valued social relations must be reflected in people's minds, in their dispositional knowing and willing.

Although collaborators are usually contemporaries, they need not be. Just as researchers can co-operate with colleagues at another institution, they can also co-operate with colleagues of another generation. They can intend their work as part of an ongoing quest in which successive cohorts of scientists aim at a discovery of truth that will belong to them all. Similarly, a chess player can intend partly to sustain a tradition of play that runs from past masters to ones unborn. Finally, a person can see his whole life as part of an overall human enterprise, to which different generations contribute and which his own acts do a little to advance. As before, this fitting has no value if it is not intended or if its other parts are not achieved. But people who connect themselves to a larger project, by continuing some human tradition, achieve a distinctive rational excellence.

As within a life (9.4.1), these claims have conservative implications. If earlier generations have conceived a general human project, we give our own lives more value by continuing this project. (We also give their lives more value, by ensuring the success of some of their aims [8.4.3].) But the implications are defeasible and can be outweighed if change is on other grounds desirable. What generality supports is a moderate conservatism, giving some weight to tradition but not always blocking innovations.

10.2.3

Goods such as political action and co-operation help make Aristotelian perfection a co-operative value, one whose achievement by one person often promotes rather than hinders it in others (5.4.2). That perfection is co-operative is already to some extent the case. Knowledge in one person often stimulates it in others; games such as chess allow two players to exercise skill simultaneously. But the point is strengthened by our recent deductions from extent. If it is intrinsically good to administer other people, I can improve my own life by arranging social conditions that benefit them. If collaboration benefits all members of a group, some projects that are good for me are also good for others. At the most general level, extent makes beneficence—or the active pursuit of others' perfection—intrinsically good. Someone who seeks the greatest perfection of all has broader ends, and so lives better himself, than someone who is merely self-concerned. Far from necessarily involving self-sacrifice, his altruism is itself a perfection.

These last claims should not be exaggerated. Although co-operation and beneficence are perfectionist goods, they are not the only such goods, and they can sometimes be outweighed by others. What is more, the measure that values beneficence also finds intrinsic value in actions antagonistic to others: There is also extent in ends that involve harming many people or thwarting their plans. Here we face starkly a consequence that has been with us from the start. If Aristotelian perfection consists in exercising rationality, and people can exercise rationality while harming others, they can achieve perfection while harming others (2.3.2). More specifically, they can achieve perfection by using complex means to carry out a long-term plan aimed at harming many people. It does not follow that this harming is all-things-considered right: On the contrary, given an agent-neutral structure (5.2.3), it is almost always morally wrong. But on the generality account there is some intrinsic value in harming others, and this means we must moderate our claims about co-operation. A person who benefits or co-operates with others may not always do the one thing that is best for his own perfection, but he often does something close to what is best and so helps others without significant costs to himself. Perfection is not absolutely co-operative (5.4.2), but benefiting others is often one good way of living well oneself.

10.3 Love and Friendship

If co-operation and beneficence are intrinsically good, so are those relationships where most of us realize them best, namely, love and friendship. These relationships are often instrumentally valuable, providing emotional support that can stimulate both partners to their highest achievements, but they also matter intrinsically. Although no aspect of quality appears to the highest degree in love and friendship, enough are present to make these relationships important sources of excellence.

10.3.1

In the first place, love and friendship extend one's concern beyond the self to embrace another person or, in a family, several persons. They create common goals for a partnership and co-ordinated acts to achieve them. Mere well-wishing from a distance is not valuable, because it is not active. Nor is love good that leads to one-sided self-sacrifice: Aiming at states of one other person is no better than aiming just at one's own. What makes friendship specially valuable is the forming and carrying out of joint projects. Then a pair's aims stretch across persons, uniting them in a single intended and accomplished state. At its simplest, this can mean expressing a common style in decorating an apartment or mastering together a skill like canoeing. More grandly, it can mean intertwining careers, as when married scientists pursue joint research. At all levels, however, the first merit of friendship is that it directs action at overarching states involving more than one person.

This increase in extent may seem small beside those in politics or management, but there is more. The co-operative goals of friends are usually long-term, involving states of both persons at all times in the future. Friends act together not just for the moment, but over many years, and this point holds generally. Most of us care just for the present states of our casual acquaintances or of people involved with us in large-scale—but for that reason often short-lived—collaborations. Only with our friends do we form, or adopt from them, many goals for their lives as a whole, and only with friends are we able to promote these goals effectively. This builds into a friendship the kind of temporally extended concern that appears in one's own life as prudence, and it also gives it structure. Over time a friendship becomes an organized project, with many particular ends serving its larger defining goals. There are trivial ends such as shopping and engrossing ones such as raising a child, but all fit into a single hierarchy and have the added value that all such fitting brings.

We normally distinguish friendship from less intimate long-term relationships, such as that between a salesperson and a regular customer. This distinction, too, is captured by our account. A salesperson may have a temporally extended goal involving her customer, for example, that they keep making mutually profitable deals. If the relationship is only business, however, she will intend the other's part of the goal only instrumentally: she will intend the other's making profits only as a means to her own profits. Because of this mere instrumentality, the salesperson's hierarchy contains less dominance than it might (9.5.4), but a friend's hierarchy is more complex. In aiming at states of someone she loves, a friend intends these states partly for themselves, as ends she wants for their own sakes. This makes her ends doubly intended and gives them added dominance.

Perhaps the greatest perfection in friendship is how friends interact at particular moments. Knowing each other intimately as they do, friends can engage in a subtle emotional interplay. They can discriminate finely between each other's moods and desires and can meet those moods and desires with exactly tuned responses. Their interaction can be nuanced as strangers' is not. This precision of feeling has the same value as precision in scientific theories or games of chess (9.3.2, 9.5.1). Sensing a need, a lover provides not just comforting, but comforting about this specific worry in this way. Giving pleasure, the lover offers not some nice thing or other, but this special caress or this remark in this tone. In Nozick's phrase, the behaviour is "contoured to" the other's desires, fitting them not just generally but in all their particularity. It can modulate to meet small alterations in the person loved.[5] Through intimacy the interaction of friends becomes suffused with particular knowledge and can achieve precisely intended effects.

10.3.2

The various Aristotelian virtues are present to a higher degree in some friendships than in others. In particular, they are present more in carefully developed, enduring friendships than in ones that are shallow or short-lived. Perfectionism does not condemn brief sexual liaisons, for they contain nothing intrinsically evil, or oppose those encounters where strangers meet, share intimate secrets for a few intense hours or days, and then part. For Aristotelian perfectionism, however, a life containing only these encounters lacks the greatest goods in friendship. It lacks the specifically rational virtues of attachments that are stable, knowing, and complex. For me, although perhaps not for every reader, this consequence is attractive. And it shows that our account's valuing of friendship is no mere platitude. The generality measures point to specific good-making features of friendship and make specific claims about an ideal personal life.

The Aristotelian virtues can also be present in relationships very different from friendship. Someone who is consumed by hatred can intend another's suffering as an intrinsically chosen part of a larger end, and intend it for many years. And just as love and sympathy can be contoured to another's particularity, so can cruelty. The most delicate torture, too, is fitted to the victim's exact desires, doing precisely what will cause the keenest suffering. These antagonistic relationships are often difficult to carry through successfully, given the inevitable resistance of their victims, and even when they are not, they are forbidden for the harm they do to others (5.2.3). In eschewing them for friendship, we not only seek their virtues where we are more likely to attain them but also do so in a way consistent with our other-regarding duties. In fact, in seeking friendship we again do what increases another's perfection without sacrificing—but often increasing—our own.

10.3.3

Let us draw together the various implications of extent and dominance. On the theoretical side, the two generality measures value knowledge that runs from comprehensive laws down to particular explanations, with precise understanding at every level (9.3). On the practical side, they favour a unified life with long-term goals, especially if these goals require varied subgoals (9.4), and particular activities that are complex, intricate, and challenging (9.5). They recommend political activity (10.1) and co-operation, either face to face or at a distance (10.2), and the nuanced interactions of friendship (10.3). These last goods exhibit sophisticated rationality in the special realm of interpersonal relations.

These consequences are, first, diverse. On the Aristotelian view, there is not one narrowly defined intrinsic good, but many goods, and for different people the best choice among these goods will be different. For some the best life will concentrate on knowledge; for others, on personal relations. For some it will include political activity; for others, solitary devotion to a challenging pursuit. Even within the different goods there will be choices. Many different unified lives can be lived, organized around different goals, and there are many different complex activities: chess, gardening, hockey, poetry—a whole range, matched by an equal range of intrinsically valuable talents.

At the same time, the consequences are not unconnected. The different Aristotelian goods are all goods for the same reason: They all involve sophisticated rationality, or generality in one's beliefs and intentions. This is a great merit of a formal approach to quality: While acknowledging the plurality of human goods, it gives a unified account of their value. While allowing many routes to excellence, it gives a single explanation why they lead to the same destination.

10.4 Generality: Objections

Despite its attractions, the generality account of quality may invite certain objections. Let us discuss the most important.

10.4.1

The account's most controversial claim is that there can be intrinsic value in acts that harm many people by diminishing their perfection or, at the extreme, killing them. This claim, it will be argued, is counter-intuitive: Such acts should make the agent's life worse, not better. Moreover, some may argue, the claim mistakes the general character of practical perfection. Just as the formal object of theoretical reason is the true, that of practical reason is the good; practical reason cannot be realized by acts that accomplish evil. Finally, the claim destroys the parallel between theoretical and practical perfection. In forming beliefs we are constrained by something outside us: We must hold true what is true independently of our beliefs. If practical perfection is a parallel good, it must involve a similar constraint, so intending what is good is good and intending evil is evil.

It would be intuitively appealing to judge acts of harming as intrinsically evil, and this is possible in a merely broad perfectionism.[6] But it is not possible in narrow perfectionism: In the sense of "rationality" in which humans are essentially rational, rationality is not tied more to moral than to immoral acts (2.3.2). Narrow perfectionism must affirm that acts of harming can be good, but this claim can be defended.

In claiming that massive harming can be intrinsically good, narrow perfectionism does not say the harming is, all things considered, right. On the contrary, the act is hugely instrumentally evil and therefore utterly morally wrong. The theory can appeal to these facts in explaining our tendency to view the harming as intrinsically evil: We transfer the massive evil in the act's effects to the act itself. This explanation is confirmed if we try to abstract from the act's effects on others. Considered just in itself, large-scale harming can seem impressive—horribly impressive, but nonetheless impressive. It massively imposes will on the world: a Marxian-Nietzschean value. Or consider the harming in a context where its effects on others do not matter morally, for example, a maximax context (6.2). When Nietzsche says the existence of a Napoleon would justify the collapse of a whole civilization,[7] we do not think his example is ill-chosen. Napoleon disrupted millions of lives and caused the deaths of thousands of people, but if only his achievements counted, as they would given maximax, they would surely be substantial.

The deeper claims about practical perfection can also be answered. The formal object of practical reason is not the good but success, which involves, as truth does, a correspondence between one's mind and the outside world (8.4.1). Of course, for both theoretical and practical perfection there are considerations of quality, which make some truths more worth knowing and some ends more worth bringing about. But the generality account characterizes the two kinds of quality in parallel, using the same formal measures, and this gives rise in each case to a parallel arbitrariness about the content on either side of the high-quality relation. If the world contains *p,* we must believe *p,* but if it contained *not-p,* we would have the same reason to believe that. What matters is not the content of our belief but its matching the world, whatever the world contains. Likewise, for practical perfection what matters, quality aside, is just the world's matching our aim, whatever aim we had. Just as there are no intrinsically favoured belief contents, so there are no intrinsically favoured ends, just the same requirement for a relation with formally high-quality terms.

10.4.2

A second objection concerns the measure of dominance. Because dominance values complex hierarchies of intentions, it also values inefficiency. It prefers roundabout routes over direct routes. Is this preference not counter-intuitive, especially in a theory that claims to value rationality?

Dominance does value inefficiency, but not when doing so would be most objectionable. If inefficient means prevent us from achieving a final goal or make our belief that we will do so unjustified, they have costs on the dimension of number (8.2). Even if we do achieve our goal, an inefficient process has elements that do not contribute to its success: They are willed as means to an end but do not help to bring the end about. These means do not increase the end's dominance (9.2.1). Finally, inefficiency is often costly in time and resources, and so hampers other perfections. For a fair test of dominance, we must imagine an inefficient activity that achieves its goal, involves only means that contribute to this goal, and has enough intrinsic value that its ill effects on other goods are outweighed. This is precisely the kind of inefficiency we should think is valuable. If Edmund Hillary wanted to stand atop Everest, climbing was inefficient compared to chartering a helicopter, yet this hardly diminishes his feat. Nor should Jack Nicklaus worry that his golf score would be lower if he walked down the fairway and dropped his ball in the hole by hand. Games by definition forbid the most efficient means to certain goals, yet playing them can be deeply valuable. If this feature does not undermine their value, it is no blanket objection to dominance that it sometimes values inefficiency.

Although this provides some defence of dominance, the objection about efficiency can be taken further. Often we do not just disvalue inefficiency, but positively value its opposite: We find positive worth in simplicity, in the lack of parts of an elegant mathematical proof or line drawing. Does dominance not exclude this value?

Not everything that is simple is valuable for being so. A Picasso line drawing may impress us with its economy, but an aimless squiggle does not. We do not call a proof elegant if its conclusion is banal, or writing concise if it has no ideas to express: There is a difference between elegance and primitivism. These points suggest the following idea: A procedure is elegant when its overriding goal requires several subordinate goals but they are all achieved by one or a few goals subordinate to them. This occurs when an agent's goals are structured as in Figure 10.1: The Picasso drawing conveys a whole character in a few lines; the proof establishes

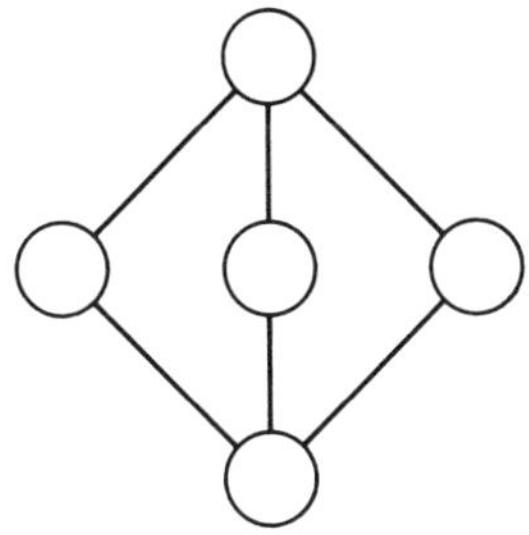

Figure 10.1

several lemmas at once. In both cases the elegance rests on an initial complexity, on the fact that several subordinate tasks are completed in one go.

If this idea is sound, we can capture the value of elegance by amending the measure of dominance to give extra credit to an end for simultaneously bringing about several other ends above it in a hierarchy. Thus, in Figure 10.1 the bottom-row end can have an extra three units of dominance for bringing about the three ends immediately above it, and it may also contribute more dominance to the ends above it. Working this proposal out in detail would involve deciding what conditions an end must meet before its accomplishing several others adds to its dominance, and also settling some questions about weighting. We intuitively value elegance, but we also value some activities for their complexity. Combining these two intuitions in one view would be delicate, but we can see in outline how a view that values complexity can be extended to value elegance and simplicity.

10.4.3

A more general objection claims that in measuring only rationality the account misses the value in human emotion. The initial response to this objection should be clear. Emotions involve beliefs and aims, and when they are better justified or more sophisticated the feelings they help constitute are by perfectionist standards good. Not all emotions embody Aristotelian quality. Unreciprocated infatuation is not good, nor are paralyzing fear and anger. (Someone prone to these states should perhaps experience them for the sake of self-knowledge, but in themselves they lack worth.) Crude emotions—ones whose expression is chaotic or undirected—lack the excellence of feelings that are structured and precise. The latter are our most sophisticated emotions, and on any view ought surely to be our best.

It may be replied, however, that these arguments still do not value emotion directly. If someone feels a nuanced sympathy for another's suffering, perfectionism will credit the knowledge displayed, but it could co-exist with indifference or even hostility. Perfectionism will also credit any actions that issue from the sympathy, but what if none do? What if the person merely feels sympathy but cannot act from it? Our theory could value mere sympathy by changing its view of number and making desires, not intentions, the basic units for practice (8.1.3). I have resisted this move, holding with Aristotle that emotions count for practical perfection only when they express themselves in action. It makes "no small difference whether we place the chief good in possession or in use, in state of mind or in activity,"[8] and I have placed the good in activity. A more inclusive view is possible, one that values feeling directly. In my view, however, the best perfectionism counts only the ends we actively intend and, given a success view of number, achieve (8.5).

10.4.4

A related issue concerns spontaneity. It may be objected that in valuing hierarchies of intentions our account leaves no room for unplanned, spontaneous action. Yet, is some spontaneity not essential for a valuable life?[9]

This objection errs, first, by treating one Aristotelian good, a unified life, as if it were the only such good. Imagine that certain experiences cannot be planned: They either just come upon one or do not come at all. Imagine also that the experiences have high Aristotelian worth. Then there is a conflict between two applications of the generality account. To live a fully organized life we must renounce the experiences, whereas to enjoy the experiences we must abandon some planning. Nothing requires that in this conflict the value of organizing should always win, and thus nothing requires that good lives always lack spontaneity.

Consider an artist's life, for example, Bob Dylan's. By moving chaotically among different musical styles and political and religious beliefs, Dylan's life may lack some overall unity. But this chaos may be necessary for him to write his greatest songs. At each time he may need a new passion to throw himself into, without thoughts for the past or future. If so, the best life possible for Dylan, given his talents and temperament, is not the most unified.

Nor must there always be conflict between these values. As Rawls points out, even worked-out life plans become less specific for the more distant future. They sketch the outlines of one's future action but leave the details to be filled in later.[10] Given life subordination in the calculation of dominance (9.2.3), this is not by perfectionist lights a failing. If an end's worth depends on the number of those subordinate to it in one's life as a whole, its present dominance is not diminished if we will choose our future means to it later.[11] And when we come to choose these means, we can be as spontaneous as we please. If our overall plan permits different elaborations, we can choose whichever among them strikes us as most appropriate when the moment arrives.

Sometimes there is even mutual support between the values. Return to the valuable experiences that cannot be planned and imagine someone who has them. Her lack of planning is necessary for the experiences and, if intentional, increases their quality: it adds to a small hierarchy around them. People often joke about "organizing spontaneity," but this is something we all do. We travel to foreign countries so we will be forced to make unarranged decisions. We know our time with friends will go worse if we try planning it in advance, so we avoid such planning. This avoiding may be habitual, but it is also intentional and a means to a successful goal. As such it adds to number and, on the quality side, to dominance.

These points do not apply to all spontaneous action, and Aristotelian perfectionism does not value everything that is unplanned. It sees no merit in sudden weakness of will or in the "spontaneity" of a life that flits among different passive experiences. Why should it? If spontaneity does not fill in an existing life plan or promote further goods such as art or friendship, why should anyone think it good?

10.4.5

A final objection extends the complaint that agent-neutral moralities, by requiring an impartial concern for all persons, give insufficient weight to particular emotional attachments.[12] If sound, this complaint applies especially to Aristotelian perfectionism. Given its measure of extent, this theory holds that a person who pursues the excellence of everyone not only does more of what is right than one who helps only her intimates but also achieves more intrinsic good. So impartiality is doubly required. The theory gives arguments for the value of friendship (10.3), but these, the objection claims, treat friendship only as a second best. What the arguments value in particular relationships appears more fully in impersonal beneficence, so our attachments to particular friends reflect only our incapacity for more generalized emotions.

But is friendship really more than a second best? Imagine a person with a small amount of concern for strangers and a great deal for her friends. It would clearly be undesirable if she abandoned her love for her friends and came to care for no one more than a small amount. Would it also be bad if she came to care for everyone a great deal? Would her life be worse if she increased her concern for strangers to the level once reserved for her friends? I cannot see why. If she was sufficiently sensitive to have knowing exchanges with everyone and to further all their long-term goals, this would surely be good rather than bad. In everyday life we have our most valuable interactions with our friends, because our greater knowledge of them permits this. But our attachments do not matter as particular attachments, and it is not wrong to value them only as means to relationships that could in principle be impersonal.

10.5 Generality: The Tradition

Alongside its moral appeal, the generality account has a distinguished history. Perfectionists with very different-sounding theories of human nature often elaborate their values by using something like extent and dominance. Consider, for example, Aristotle, Leibniz, Nietzsche, and Bradley.

10.5.1

Aristotle does not use only generality measures of quality. He thinks the best knowledge is of certain divine objects and the best acts pursue morally worthy ends. His view of quality is therefore partly material, whereas ours has been purely formal. Much in his theory of human nature turns on generality, however, and he often makes the particular moral claims that our two measures support.

Aristotle values rationality partly as something distinctive of humans (2.1.1), and he is quite clear what makes it distinctive. Reason, he says, is the faculty that knows universals, whereas the perception we share with animals grasps only particulars.[13] Rationality is therefore defined in terms of the generality of mental contents. This suggests that a human life will be more perfect as its beliefs and aims are more general, and Aristotle often makes this claim. He thinks everyone should have a single end to aim at in life, because "not to have one's life organized in view of some end is a mark of much folly,"[14] and he calls only acts directed to such an end truly "chosen."[15] He believes that good activities are better when they are more difficult[16] and treats the practical wisdom of the politician, who deals with many people, as of a higher order than that of ordinary citizens.[17] Even on the theoretical side, he could justify his preference for knowledge of divine objects without abandoning formal measures. The eternal, unchanging beings are also the causes of all other beings,[18] and knowledge of them is therefore most explanatory. The hierarchical importance of this knowledge, not the intrinsic merit of its objects, could justify its primacy.

10.5.2

Leibniz equates our essence with clear and distinct perception of the universe, so a soul "has perfection in proportion to its distinct perceptions."[19] But the best perceptions, for Leibniz, are the most general:

> One must consider in God a certain more general and comprehensive will, which he has regarding the whole order of the universe, since the universe is like a totality which God penetrates with a single view. . . . Indeed the wiser one is, the fewer separate acts of will one has and the more one's views and acts of will are comprehensive and linked together. And each particular act of will contains a connexion with all the others, so that they may be harmonized to the highest possible degree.[20]

10.5.3

Nietzsche, too, values generality. In his early works, before he has discovered the will to power, he ties excellence to the presence in a life of unifying ends:

> *One thing is needful.*—To "give style" to one's character—a great and rare art! It is practised by those who survey all the strengths and weaknesses of their nature and fit them into an artistic plan until every one of them appears as art and reason and even weaknesses delight the eye. . . . In the end, when the work is finished, it becomes evident how the constraint of a single taste governed and formed everything large and small. Whether this taste was good or bad is less important than one might suppose, if only it was a single taste.[21]

When the will to power does appear, it is characterized in terms of the same unification:

> *Weakness of will:* that is a metaphor that can prove misleading. For there is no will, and consequently neither a strong nor a weak will. The multitude and disgregation of impulses and the lack of any systematic order among them result in a "weak will"; their co-ordination under a single predominant impulse results in a "strong will": in the first case it is the oscillation and the lack of gravity; in the latter, the precision and clarity of direction.[22]

> What is essential "in heaven and on earth" seems to be, to say it once more, that there should be *obedience* over a long period of time and in a *single* direction: given that, something always develops, for whose sake it is worth while to live on earth; for example, virtue, art, music, dance, reason, spirituality.[23]

The reference to a "long period" suggests extent across times, and Nietzsche often says the powerful achieve enduring states. He says a great individual "can extend his will across great stretches of his life" and looks forward to a new caste that will rule Europe with "a long, terrible will of its own that would be able to cast its goals millennia hence."[24] There are also hints of extent across persons. The will or force directed inwards by an artist or thinker "is the same active force that is at work on a grander scale in those artists of violence and organizers who build states." In the artist, however, where it works "internally, on a smaller and pettier scale," its object is "man himself, his whole ancient animal self—and *not,* as in that greater and more obvious phenomenon, some *other* man, *other* men."[25]

10.5.4

Last is Bradley, representing Idealist perfectionism. At the deepest level, Bradley equates perfection with realizing one's identity with the Absolute, but his more specific claims turn on generality. To estimate the value of an appearance, or what he calls its "reality" (3.1.1),

> We have, on the one hand, to consider its harmoniousness. We have to ask, that is, how far, before its contents can take their place as an adjective of the Real, they would require re-arrangement. We have to enquire how far, in other words, these contents are, or are not, self-consistent and systematic. And then, on the other hand, we must have regard to the extent of time, or space, or both which our appearance occupies. Other things being equal, whatever spreads more widely in space, or again lasts longer in time, is more real.[26]

These two tests of "harmoniousness" and "extent" apply equally to theoretical and practical perfection, for "the practical standard seems to be the same as what is used for theory."[27] The best knowledge "must exhibit the mark of internal harmony, or, again, the mark of expansion and all-inclusiveness."[28] And in action,

> To reduce the raw material of one's nature to the highest degree of system, and to use every element from whatever source as a subordinate means to this object, is certainly one genuine view of goodness. On the other hand, to widen as far as possible the end to be pursued, and to realize this through the distraction and dissipation of one's individuality, is certainly also good. An individual system, aimed at in one's self, and again the subordination of one's own development to a wide-embracing end, are each an aspect of the moral principle.[29]

Bradley's explicit talk of generality matches much in other Idealist perfectionists. Despite their elaborate metaphysics, these perfectionists have an account of quality close to that of Aristotle, Leibniz, and Nietzsche and well suited to the idea that humans are essentially rational. They have, I would say, the standard account of quality in serious narrow perfectionism.

III

PERFECTIONISM AND POLITICS

11

Liberty

To this point we have considered perfectionism as a personal morality, which says how private individuals should act, but the theory naturally extends itself to politics. Given its agent-neutral maximizing structure, it supports the following standard for political evaluation: The best political act, institution, or government is that which most promotes the perfection of all humans. This standard can be used to judge governments' external behaviour, for example, to condemn them for starting aggressive wars, but its more common application is internal. Here it says the best government most promotes the perfection of all its citizens. Often (although not necessarily) this means that the best government is explicitly committed to perfectionist values, or to promoting a perfectionist vision of the good.

This political standard is affirmed by many perfectionists. Aristotle, for example, says the state exists "for the sake of a good life," by which he means a perfect life, and that under the best government all citizens act best;[1] similar statements appear in Aquinas, Green, and Rashdall. But its implications for politics are a common source of resistance to perfectionism. In particular, critics worry that perfectionism is hostile to the modern political values of *liberty* and *equality*. Because it thinks some lives are better than others, they argue, regardless of whether people want or would choose them, it favours state coercion to force people into excellence. Also because it thinks some lives are better, perfectionism wants the bulk of resources distributed to those who can lead such lives. Instead of equal shares, it wants the material conditions for perfection confined to a small elite, for whose benefit the rest of the population must labour.

These charges take colour from parts of the perfectionist tradition. Plato and Aristotle want governments to mould their citizens' characters, with no restraints on how this moulding is done, and think education and political power are wasted on those who perform manual labour. Nietzsche goes further, exulting in a vision of aristocrats who will "use the great mass of people as their tools."[2] As I will argue, however, the charges do not apply to all perfectionisms, and especially not to the perfectionism we have developed. This theory already contains elements that support individual liberty and a roughly equal division of resources, and in this it conforms to parts of the perfectionist tradition. Alongside the perfectionism of Plato and Nietzsche is that of Marx, who wants the state to wither away so humans can enjoy the "free play of their own physical and mental powers"[3] and who favours distributing resources in accordance with needs. Less radically, the Idealist period in

Britain saw a broadly based movement, inspired initially by Green, that derived a welfare-state liberalism from perfectionist principles. During this period modern egalitarian liberalism, as against nineteenth-century laissez-faire, first developed, and it did so largely under perfectionist guidance. Taking liberty in this chapter and equality in the next two, I will argue that some egalitarian liberalism—perhaps socialist, perhaps social-democratic—is the natural consequence of Aristotelian perfectionism and thus of perfectionism in its most attractive form.

11.1 Autonomy as a Perfection

Philosophical discussions of liberty have often focussed on the harm principle, or *liberty principle.* Stated initially by Humboldt and John Stuart Mill, this principle says that the state must never coerce its citizens except to prevent them from harming others or interfering with others' liberty. This principle defines classical liberalism and is what perfectionism is most often said to violate. It is therefore our subject in this chapter. We will ask whether perfectionism can affirm a version of the liberty principle or, on the contrary, favours state interference with citizens' self-regarding action.

11.1.1

Let us start by considering perfectionism just in the broad sense, without foundations in human nature (1.2). Broad perfectionism can most easily affirm a liberty principle by treating *autonomy,* or free choice from many life options, as itself an intrinsic good. If self-determination is itself a perfection, any restrictions on it are prima facie objectionable.

Humboldt and Mill both take this broadly perfectionist line. Humboldt says it is part of human perfection that one "strives to develop himself from his own inmost nature, and for his own sake."[4] For him the chief requirements for perfection are therefore "freedom" and, what he calls "intimately connected" with freedom, a "variety of situations."[5] Mill says the qualities that are "the distinctive endowment of a human being" are exercised "only in making a choice."[6] To encourage such choices he recommends, again, no interference in the private realm and "experiments of living"[7] to make different life options vivid for all.

Humboldt and Mill do not affirm just some liberty principle; they affirm an absolute principle. They say that in actual conditions the state must never interfere in citizens' private lives. Can this absolute constraint be derived from a perfectionist valuing of autonomy?

Even considering only the value of autonomy, it cannot. Sometimes restricting a person's autonomy now will do more to increase her autonomy in the future, by giving her more options in the future or a greater capacity to choose autonomously among them. Humboldt and Mill may believe that this situation arises infrequently and that governments are especially inept at recognizing it. Nonetheless, it is one bar to a simple argument from the value of autonomy to an absolute liberty principle.

A more serious impediment comes from the recognition of perfections other than autonomy. No plausible value theory can treat free choice as the only intrinsic good.

It must acknowledge some other goods, so that, for example, freely chosen creativity is better than freely chosen idleness, and autonomous knowledge is better than autonomous ignorance. Does this not reopen the door to interference? If there are goods other than autonomy, may sufficient increases in them not outweigh any loss in autonomy?

They may do so if autonomy is just one good among others, to be weighed against them in standard fashion. But there are other possibities. A broad perfectionism can treat autonomy as lexically prior to other perfections (7.1.1), so that even small losses in it outweigh large gains in those others. More moderately, perfectionism can treat some minimum of autonomy, one involving reasonable self-determination, as lexically prior, so that losses that take one below this minimum (although not other losses of autonomy) outweigh large gains in other perfections.[8] Finally, the theory can treat autonomy, not as a good among others, but as a condition of goods. Then states and activities that would have value if they were chosen freely have none if they are coerced. Autonomy is not a competitor with other goods, but a condition of their worth.

These three views all limit the possibility that losses in autonomy will be outweighed by increases in other goods, but only the first—full lexical priority—does so entirely. The second and third views exclude the most serious restriction on liberty: forcing people into a single best activity. But they do not exclude the milder illiberality of merely forbidding some worst activity. Imagine that ten self-regarding acts are ranked in value from 1 to 10. An absolute liberty principle condemns laws that forbid just the tenth-ranked option, but these two views do not. Such laws leave people free to choose among the nine remaining options, thus leaving them reasonable autonomy and any autonomy that could plausibly be a condition of goods. Their loss of liberty is not of the kind that, according to the two views, has special political significance.

Whatever their precise implications, none of the three views is plausible as a view about the good. Imagine that Mozart was as a young boy forced into music, so his life lacked reasonable autonomy and the autonomy that may be a condition of goods. All three views imply that Mozart's life, despite its great musical achievements, contained less perfection than if he had been given freedom in his youth and had autonomously chosen a life of suntanning. This claim is not one that, if we think seriously about the good, we can make. Even if autonomy has some value, it cannot have so much as to outweigh all Mozart's music.

A plausible broad perfectionism, then, can treat autonomy only as one good among others, which may sometimes be outweighed. It therefore cannot endorse an absolute liberty principle, but it can endorse a non-absolute principle. If free choice is intrinsically good, any restriction on it threatens some perfectionist cost and is therefore prima facie objectionable. Other things equal, the state should not interfere with liberty except to protect the greater liberty of others.

11.1.2

It is important that autonomy can be added to a list of merely broad perfections, but it would be more impressive to justify its place there. In our Aristotelian perfectionism, we would need to show that autonomy develops human nature by exercising sophisticated rationality. Can autonomy be valued as an exercise of reason?

There is an obvious connection between autonomy and rational deliberation. Someone with many life options can reflect upon these options and in so doing exercise his rational powers. He can weigh their respective merits and defects and reach a reasoned judgement about them.

This connection is not, however, quite what we want. As described previously, deliberation is an intellectual activity that does not require options actually to be open. Someone who has no free choice—a slave at the extreme—can still evaluate possibilities. He can ask which career would be best for him if he were able to choose careers and can arrive, in principle at least, at the same intellectual conclusion as someone who has many careers open. Of course, a slave is unlikely to go through this process. People rarely deliberate about options they cannot choose, and this is an important instrumental argument for liberty. By giving its citizens many alternatives, a free society encourages reasoning that would have no practical point if options were closed. But the argument is instrumental and does not yet give us rational values present in the act of choice itself.

To locate these values, we must examine choice more closely. Imagine that one person chooses life *a* from among ten life options, while another person has only life *a* available. It may be true of each that she has made *a* the case and is in that sense responsible for it. But there is an important difference between them. The first or autonomous person has also made nine alternatives to *a* not the case: If her options included *b, c,* and *d,* she is responsible for not-*b*, not-*c*, and not-*d*. The second person did not have this further effect. Her non-realization of *b, c,* and *d* is due, not to her, but to nature or to whatever person limited her options. This difference is important because practical perfection involves agency: It involves expressing intentions in the world and determining what it does and does not contain. The autonomous agent, by virtue of her autonomy, more fully realizes this ideal. When she makes choices, she has two effects: realizing some options and blocking others. She has a more extensive causal efficacy than someone who lacks options, and a higher score on the practical dimension of number. By letting her determine what she does not do as well as what she does, her autonomy makes her more widely active and more practically efficacious.[9]

To have value, this efficacy requires more than just the availability of options. A person must know about the options or she cannot intend their non-realization. Moreover, she must make in the fullest sense a choice among them: a choice that is for one option and against others, so her rejection of the others appears in her mind. This does not always happen in intentional action. Someone who is driven by an obsession may know that alternatives to her act are available, but her intentions do not reflect this. They go blindly for *b,* if *b* is what she does, without preferring *b* to other options. (A strong claim is that the obsessed person intends only *b,* without rejecting anything; a weaker claim is that she rejects only the vague alternative not-*b*. Either way there is not the rejection of individually discriminated options that on my view increases agency.) The same is true of someone who is weak-willed. If he acted on an all-things-considered desire for *a,* he would prefer *a* to other options. But he does not act weakly in preference to other options; he just acts weakly. If autonomy involves successfully realizing many intentions, it requires choice in the fullest sense: a simultaneous realization of some possibilities and rejection of others, with one's rejection of the others reflected in one's will.

Autonomy realizes agency even more so when it follows deliberation. (We can now take up the earlier suggestion about practical reasoning.) Someone who reflects on her options may find that *a* has the most of desirable property *F*, *b* has defect *G*, and so on. When she chooses, therefore, she can intend not only *a* but the-act-with-the-most-*F*, and in rejecting *b* she can reject the-act-with-*G*. Her deliberative knowledge, if it guides her choice, can give her more intentions in and around her options than if she picked blindly among them. There is a further effect. If she has deliberated, she can choose *a* and reject *b* as means to a single goal, perhaps getting the-most-*F*-without-*G*. Her positive and negative intentions can form a unified means-end structure, and the more complex this structure is, the more it embodies dominance. Then we have not just autonomy but *deliberated autonomy:* free choice from a wide range of options that reflects practical reasoning about them. This kind of autonomy realizes deliberative rationality, but in a more than intellectual way. The elements it organizes are not beliefs, which are available to a slave, but intentions realized in the world. It therefore presupposes the simpler autonomy that consists in any free choice among options. Its foundation is a set of acceptances and rejections that converge on one goal, which is possible only for an agent who has many options to accept or reject.

11.1.3

An objection may be raised against this account of autonomy. If it is good to achieve additional ends, it may be argued, we should, when doing something trivial such as lifting a fork, think of the many things we are not doing and consciously reject them. We should think that we are not lifting our knife, using our other hand, and so on. Is this not absurd?

The reference to consciousness here is a red herring because ends can be willed or rejected without conscious awareness (9.5.2). Even apart from this point, the objection fails by ignoring the dimension of quality. The alternatives we could reject when we lift a fork are highly particular and, as such, of minimal worth. So if autonomously manipulating cutlery distracts us from greater goods, as it surely will if it is conscious, it is on balance wrong. The point is like one about knowing the number of redheads in Beiseker, Alberta. In itself this knowledge has (minimal) value, but someone who takes the time to acquire it will hardly be maximizing his excellence. In the same way, some gains in autonomy are too trivial to be worth seeking. What matters in Aristotelian perfectionism is the autonomous choice of encompassing ends, ones that shape a day, month, or life. In this kind of choice, we both select and reject goals of high quality and thereby realize substantial agency.

This last point is a further implication of dominance. Earlier this measure preferred autonomous choices that follow deliberation; here it and extent prefer choices, whether deliberated or not, that are of important or organizing ends. This implication fits the views of Humboldt and Mill, who care most about the freedom to fix the general shape of one's life or one's general guiding ends. There are doubtless instrumental reasons for this preference; any extrinsic benefits of free choice are surely greatest when the choice is most far-reaching. But there is also an argument about intrinsic value. Autonomous choice seems most valuable, as autonomous choice, when it is of important, organizing ends. Our generality measures explain why.

There is also a connection with Humboldt's and Mill's interest in "a variety of situations" and "experiments of living." As we have seen, autonomy requires knowledge of available options, but it seems fullest when this knowledge is most extensive, when the options are grasped not vaguely but in articulated detail. Again our generality measures explain why. Someone who lives in a varied society can choose among life forms he knows at first hand and, when he chooses, accept and reject not just vague possibilities but connected series of acts. Within a large hierarchy uniting choices and rejections, he can have many smaller hierarchies representing all the worked-out options he declines.

There is therefore an Aristotelian account of autonomy, which values most highly those choices that follow deliberation, are of organizing ends, and involve articulated knowledge. But the account has an important limitation. Because it values autonomy for the same reason it values other goods, as a realization of rationality, it cannot plausibly treat autonomy as special among goods, for example, as lexically prior to or a condition of goods. This limitation, however, is not troubling. The stronger valuings of autonomy are not intuitively plausible (11.1.1), and a perfectionism that rejects them still has at least one argument against any illiberality: If autonomy is one intrinsic good, interfering with citizens' private lives always threatens some perfectionist harm. This argument may not always be decisive, especially against the milder illiberality of merely forbidding one worst option (11.1.1). But its non-absolute character is less important when we realize that it does not operate alone. Beyond the autonomy view is a further perfectionist argument for liberty, one that would be available even if autonomy were not directly valuable or free choice itself a good.

11.2 The Asymmetry Argument

11.2.1

This further argument starts from the *asymmetry* (5.3). Because Aristotelian perfection is active and inner, it is not something individuals can often directly produce in each other. Past a point, each person's achievement of perfection must be his own. The same limitation applies to governments: They too cannot directly produce their citizens' good. They can supply necessary conditions for their citizens' perfection, or conditions that make this perfection more likely, but the sufficient conditions are beyond their power.

Illiberal action as normally envisaged tries, not to produce perfection directly, but to induce it by legal threats. But the same factors that support the asymmetry tell also against the effectiveness of such threats. Because Aristotelian perfection is largely inner, it is not open to the public inspection that effective regulation requires. How could a government be certain how nuanced a citizen's intentions were, or whether he intended his acts as parts of a unified life? Given this uncertainty, how could laws requiring such inner states be enforced? At the same time, the formality of Aristotelian perfection means that the relevant inner states are only loosely correlated with outward behaviour. The same outward acts that, approached with the right intentions, have high value can, if willed differently, embody little value. If they are chosen only as means or without any relation to broader ends, they involve little extent or dominance. So even indirect coercion, restricting outward behaviour in the hope of inducing the valued inner states, would not often be effective. People could act within its constraints yet achieve little of worth.

The asymmetry is the core of Green's argument for liberty. For Green the good is "character," and "no one can convey a good character to another. Every one must make his character for himself."[10] The same limitation applies to governments:

> It is the business of the state, not indeed to promote moral goodness, for that, from the very nature of moral goodness, it cannot do, but to maintain the condition without which a free exercise of the human faculties is impossible.[11]

Given the nature of perfection, in other words, the proper function of government "seems necessarily confined to the removal of obstacles."[12]

Green's argument recurs in many later writers. R. L. Nettleship says, "We differ from Plato and Aristotle not in our view of what is fundamentally important to the community, but in the line we draw between things with which the state can interfere to advantage, and things which it should leave alone."[13] Quoting Nettleship, Sir Ernest Barker continues:[14]

> Ethics hardly figures in our political science in the same way as in that of Aristotle. The state cannot be said to habituate its citizens actively in the ways of virtue. Once the state attempted the task in England, under the commonwealth; and it raised up in one generation a crop of imitative hypocrites, and in the next a crew of reactionary debauchees. Ethical life, we feel, is nothing without spontaneity. Automatism has no moral value; and the aim of legislation is to get rid of itself. The modern age sets itself therefore to the removal of the obstacles to a moral life.

For a generation of perfectionists after Green, the good consists, as for Plato and Aristotle, in a certain inner state of character. But there is now a conviction that this state cannot be imposed from without, and a consequent rejection of illiberal laws. This connects perfectionism with liberalism and also gives liberalism a new rationale. The liberal commitment to liberty need not rest on agnosticism about the good or on the view that only free choice is good. It can be grounded in a deep fact about human perfection: that each person's achievement of it must be largely her own.

11.2.2

Although the asymmetry begins an argument for the liberty principle, it cannot complete it. It shows that coercion does not generally promote perfection and therefore lacks a positive rationale. But this fact only makes coercion morally indifferent, and a serious defence of liberty must show that it is wrong. It must show that state interference with self-regarding action not only fails to do good but also is likely to do harm.

Green saw this need and argued that illiberal laws make people's inner states worse. The "character" he values is partly moralistic: It partly involves benefiting others from social concern or from a sense of duty. And the moralistic elements in character are undermined by legal threats: "The end consists in action proceeding from a certain disposition," and "action done from the apprehension of legal consequences does not proceed from that disposition."[15] Acts motivated by the fear of punishment are not done from a morally valuable motive.

Because our Aristotelian perfectionism is not moralistic, it cannot take this simple line of Green's. It cannot hold that any act from non-moral motives lacks intrinsic value. It can, however, point to several harms that illiberal laws can do.

First, these laws can limit people's routes to excellence. To be enforceable, the laws must restrict outward behaviour (11.2.1), but this behaviour is only loosely correlated with desirable inner states. Just as any activity can, if done for the wrong reasons, lack value, so any activity can have value. If what matters is a person's intentions, almost any act can, if approached in the right spirit, embody some perfection. Of course, there are links between behaviour and inner states. The intentions of chess players can be more sophisticated than those of checkers players because the rules of their game permit this. But the links are not exceptionless. What is for many a lesser activity can for some represent a good or even their best perfectionist option. To forbid it is to deny them something of genuine perfectionist worth.

Second, illiberal laws can make people's intentions worse. The goal of avoiding punishment is, as a goal, of limited generality. Turning as it does on harms threatening oneself in the near future, it is not usually extended in times or persons or often part of a complex hierarchy. On the Aristotelian view, acts are best when chosen as parts of an organizing plan for one's whole life or, better still, for one's community. The same holds for non-acts: Choices against a course of action are best when they are directed to a larger purpose. Legal threats undermine these connections. They introduce the fear of punishment, which can supplant other motives. Even if people choose for and against as the law requires, they can do so just to avoid punishment. If some would otherwise have acted in the same way for better reasons, there has been a perfectionist loss.

Equally important are the effects on double intending (9.5.4). Given our measure of dominance, the best acts are chosen partly for themselves rather than just as means to something beyond them. Again, legal threats undermine this attitude. Imagine that there are laws requiring chess and, because of them, people play more skilful chess than before. If some of them play only to avoid punishment, their activity can have less value than less able chess chosen partly for itself. Their individual moves are willed only as means and do not have the extra value that comes from being willed intrinsically. That activities are best when they are chosen partly for themselves is not just an Aristotelian but a generally attractive claim, and it tells further against illiberal laws. Not only can these laws not produce intrinsic choosing, but they make it less likely. By encouraging an exactly opposed motivation, they inhibit its growth.

This last argument appears, alongside remarks about formality, in Humboldt. "There is no pursuit whatever that may not be ennobling," he writes; all that matters is "the manner of its performance." Here a person's "pursuits react beneficially on his culture, so long as these, and the energies allied with them, succeed in filling and satisfying the wants of his soul; while their influence is not only less salutary, but even pernicious, when he directs his attention more to the results to which they lead, and regards the occupation itself merely as a means."[16]

These arguments about motives would have less force if excellence could be induced by habituation, so that people who first chose goods to avoid punishment came later to will them intrinsically. Then the bad effects of illiberality would be confined to a short initial period, after which the best acts would be done from the best motives.

Habituation often works with children and is therefore desirable for them. Our schools force children to experience music, literature, and athletics in the often justified hope that they will later come to choose these pursuits for themselves. The same tactic is less effective with adults, who usually have fixed values and interests and are therefore harder to lead to new ones. What is more, they tend to resent directives about their private lives and to obey them at best grudgingly: Their attitude at the start of habituation is precisely not ripe for developing intrinsic choosing. Imagine again that there are laws requiring chess. Will people be led by these laws to love chess for itself? Or will they view chess as something an authority requires, to be played because it is commanded and for no other reason? Some illiberal perfectionists, for example, Rashdall, are more sanguine about the habituation of adults and favour some laws that attempt it.[17] But my inclination is to side with Green and to say that the attempt to habituate adults into excellence is usually counter-productive. Instead of developing the best motivation, it more commonly strengthens attitudes inimical to it.

Finally, illiberal laws can undermine the general character needed for excellence. By supplanting good motives in one area, they can inhibit their development in others, and they can also diminish energy and self-direction. As Humboldt says,[18]

> The evil results of a too extensive solicitude on the part of the state, are still more strikingly shown in the suppression of all active energy, and the necessary deterioration of the moral character. . . . The man who is often led, easily becomes disposed willingly to sacrifice what remains of his capacity for spontaneous action. He fancies himself released from an anxiety which he sees transferred to other hands, and seems to himself to do enough when he looks to their leadership and follows it.

11.2.3

As so extended, the asymmetry argument combines with the autonomy view to comprise a composite perfectionist case for liberty. Government interference with self-regarding action reduces citizens' autonomy and especially their deliberated autonomy. At the same time, it rarely succeeds in promoting other perfections and can work in several ways to diminish them, by removing routes to excellence, inducing less valuable motives, and weakening self-direction. Although its elements are all prima facie, the case as a whole is impressive, and a perfectionism that accepts it can affirm a fairly strong version of the liberty principle. It may even be argued that it should affirm an absolute version. What we conclude from the perfectionist case depends on how adept we think the state is at recognizing the exceptional cases where illiberal laws are justified. If the state is skilled, it should be given rein; if it is inept, it should be held back; and if the state is very inept, the best practical policy may be for it never to interfere with self-regarding acts. Although violations of liberty can in principle be justified, they should never in practice be attempted.

I do not believe perfectionism supports this strong a conclusion. Autonomy and the asymmetry argument tell heavily against the extreme illiberality of forcing people into a single best activity, and perfectionism will rarely if ever approve this illiberality. But they have less force against merely forbidding some one worst activity, for example, forbidding the tenth-ranked of ten activities. This milder illiberality reduces autonomy only a little, removes just one route to excellence, and leaves considerable scope in the choice among the remaining options for the operation of good motives. If it steers enough people into better activities, it can be on balance justified and can be recognized by a government as justified. So although perfectionism has a strong general commitment to liberty, it may sometimes favour legal restrictions on especially bad activities. It may, for example, favour laws against addictive drugs if these drugs are very harmful to their users and very difficult to abandon once they have been tried.

This limitation does not, however, undermine our general conclusion. The common objection to perfectionism is not that it can sometimes favour restricting liberty—this is true of many moral theories—but that it does so systematically. And the objection is answered if the theory can affirm a strong although not absolute version of the liberty principle. The point should not, however, be argued just in the abstract. To illustrate the perfectionist case for liberty, let us consider two issues that have been prominent in discussions of the liberty principle: the enforcement of sexual morals and legal paternalism.

11.3 Sexual Enforcement and Paternalism

11.3.1

Aristotelian perfectionism finds many values in sexual activity. By its lights, the best sex dwells on the process of arousal rather than simply hurtling towards climax; is communicative, as partners use subtle cues to determine jointly when their excitement will be held back and when allowed to rise; and is part and chosen as part of a longer-term relation with larger concerns. Should these values be enforced by law?

The Aristotelian sexual ideal differs importantly from the views normally behind a call for the enforcement of sexual morals. These views treat some forms of sex, not just as less good than others, but as intrinsically evil. To engage in them is not just to miss out on positive value, but to realize negative value. A narrow perfectionism cannot make this claim. Because essential properties cannot be developed to negative degrees, our theory cannot admit intrinsic evils (8.1.2) and therefore lacks the strong motive other views have for wanting some sex banned. There is a further difference. The other views usually define bad sex physically, as involving partners of the same sex or a non-standard arrangement of organs. All this Aristotelian perfectionism ignores. Because our theory's values are internal, and realizable in many bodily forms, it has no interest in traditional bans on, for example, non-procreative intercourse.

Could the Aristotelian ideal not be enforced in its own way? Laws directly requiring inner states would be unenforceable, but we can think of laws forbidding, say, sex outside marriage as aimed to promote sex with knowing, long-term intentions. May these laws not be justified?

There are several objections against such laws. First, they reduce autonomy, preventing those who do choose married sex from themselves determining that they do not practise other forms. The effects on deliberated autonomy are even worse. People who have many sexual options are encouraged to reflect on the merits and defects of these options and to choose for and against them as means to more general aims. By preventing the multiple choosing-against, the laws prevent this substantial exercise of practical rationality.

Second, the laws are unlikely to have many positive effects. Because thoughtless, uncaring sex can occur inside marriage, the laws are inefficient tools for promoting the good, and they can work in several ways to hinder it. Unmarried sex is at worst a low-value amusement, and as such no more worth banning than other amusements. Moreover, sometimes it is more than an amusement. For some people, sexual intimacy can, or can only, develop outside legal forms: To confine these people to marriage is to remove what may be their best sexual option. Finally, laws restricting sex to marriage can corrupt people's attitudes. The most valuable sex involves other-directed intentions, willing another's states both for themselves and as parts of a larger whole. But people concerned to obey a law or to get away with breaking it will not look primarily at each other's responses. With some of their attention diverted to an external authority, they will have less to give each other.

It may be replied that laws confining sex to marriage have an educative function. They affirm society's commitment to a sexual ideal, they prevent people from becoming habituated into sex that contradicts this ideal, and they encourage people to learn, through experience of the best sex, to choose it for its own sake.

The state does have an educative function (11.4); the question is whether the most effective approach to education is coercive. We want people to choose the best sex from the best motives, understanding why it is good and willing it for those reasons. But then coercion is a clumsy device. Instead of encouraging rational evaluation, it implicitly tells citizens to obey unthinkingly, without considering alternatives. Far from encouraging informed decision-making, it discourages it. If people are to choose the best sex intelligently, far better to let them learn from experience, both their own and others', of its better and less good forms. If they still need guidance, far better to make it non-coercive: describing the merits of the best sex, encouraging them to choose it, but not forcing them into it or pre-empting the use of their rational powers.

11.3.2

The state practises legal paternalism when it forbids people to do in the present what will harm them in the future. According to perfectionism, the relevant harm is to their future perfection or their future capacity for perfection. Is coercion to prevent this harm permitted?

Perfectionism has no absolute objection to paternalism, and neither did Green. He supported the health and factory acts of his day, compulsory education, and even in some conditions the prohibition of alcohol. Today perfectionism may approve seatbelt legislation, compulsory medical insurance, and perhaps laws discouraging smoking. But it has serious reservations about these laws. Paternalism should be used only rarely and after a careful weighing of costs.

It is clearly desirable that people preserve their capacity for perfection, but it is best when, unforced, they do so on their own. Then they themselves determine that they do not destroy their capacities, and if they have reasons for their action, do so in a way that involves a hierarchy of intentions. Then, too, they act on the temporally extended goals and have the overall life plan that is valuable in prudence (9.4.3). In the best case of all, they choose self-preservation partly for its own sake, as an intrinsically valued exercise of self-management. Perfectionism's objection to legal paternalism is that it undermines these rational values. In a particular case, someone who would otherwise act prudently for future-directed reasons may do so just to avoid punishment. More generally, legal paternalism can create a climate where citizens look to others for guidance rather than learning to govern themselves. The last point is often made by writers on paternalism. They say that adults who are treated like children will act like children and that people who are not allowed to make mistakes cannot learn from their mistakes. What is less often recognized is the implicit perfectionism of these remarks. Self-reliance and self-management are not instrumentally necessary because the state can practise additional paternalism later. But they are attractive traits intrinsically: To prudently safeguard one's future is to realize part of a rationalist ideal. If paternalism discourages this safeguarding, it does perfectionist harm.

11.3.3

In these applications of the perfectionist case for liberty, as in the general case itself, specific features of Aristotelian perfection play a vital role. If perfection did not involve inner states, laws requiring it directly might be enforceable; if it were not defined formally, its correlation with outward behaviour might be closer and indirect coercion more promising; if it did not matter what motives people act on, the argument that threats induce bad motives could not be made. Rival perfectionist theories are imaginable in which coercion into a good life is more often justified, but our Aristotelian values were developed as those most plausible in themselves. If they support liberty, perfectionism in its most plausible form gives, if not absolute, then strong support to individual liberty.

11.4 Liberty versus Neutrality

11.4.1

Although perfectionism supports liberty, it does not go beyond it to support the stronger ideal of *state neutrality*. According to this ideal, the state must not only not coerce citizens to make them better, it must never aim, coercively or otherwise, to promote one set of values over others. It must be neutral about the good, never having as its justification for acting that some ways of life are intrinsically preferable to others.[19]

Neutrality is not a traditional liberal ideal, for it is rejected by Mill: He thinks a person's choosing badly, although no reason to coerce her, does justify "remonstrating" and "reasoning" with her.[20] Nor is neutrality supported by our perfectionist arguments. These arguments tell strongly against coercing citizens into the good, but they do not have the same force against non-coercively promoting the good. Some such promotion can have bad effects: Offering rewards for excellence can induce bad motives just as much as can threatening punishments. But many of its forms do not diminish citizens' autonomy, reduce their routes to excellence, or undermine valuable inner states. There is much non-coercive promotion of the good that perfectionism approves.

The state can, first, promote perfection through its education system. Its schools can teach students about the natural world and the history of their culture, in part because knowing these subjects is intrinsically good. They can also introduce students to literature, music, and athletics. The schools' efforts here will not be undiscriminating; there will not be courses exposing students to drug-taking or professional wrestling. The education system will lay the foundations for valuable activities, not for ones of minimal worth.

The state can continue by educating adults. The Canadian government sponsors advertisements encouraging citizens to participate in sports and other physical activity, in part because such activity is intrinsically good. Similar publicity for other perfections could, if free of non-rational persuasion, lead people in valuable directions without bad effects.

The state can also subsidize valuable activities. Some people's best talent may be for a rare perfection, one that few people practise. Their limited numbers may make this perfection more expensive than others that are intrinsically no better, but whose many practitioners permit economies of scale to be realized. Here the state can subsidize the rare activity, giving those drawn to it the same chance at a good life as the majority. Of course, the costs of subsidization must not be too great, but this condition is often met. When it is, subsidization encourages perfection in the minority, by making a good option more available, but does not coerce them into anything. (For the objection that subsidization coerces those who are taxed to finance it, see 13.3.2.)

The state may have other reasons for subsidization. There may be valuable activities that people cannot appreciate when they are young and will not try as adults if the initial costs of doing so are high. Here state aid, by lowering these costs, can help more people discover a valuable option. The state may also subsidize activities whose value people already know. It may believe that, although citizens have some tendency to seek out valuable activities, they have other, less desirable impulses, for example, to engage in consumption or passive amusement. It may fear that if the best activities cost their full market price, many who would choose them at lower prices will prefer something less valuable. It may therefore subsidize the activities to help more people's better tendencies actually guide their conduct.

These various justifications come together in a policy of government support for the arts, for example, for performing arts companies that tour across a country. By supporting these companies and not professional wrestling, the government affirms its belief that the arts are more valuable than professional wrestling and invites citizens to reflect whether this does not match their own convictions. Its support may permit some arts, for example, opera, to survive that on their own would be unsustainable, and takes them to places, such as smaller towns, that could not otherwise experience them. Finally, by lowering ticket prices, state support increases the number of people who attend the performing arts, both for the first time and regularly. The support does not force anyone to attend; they can watch wrestling if they prefer. It merely makes a valuable option more available and easier to choose from valuable motives.

11.4.2

This non-coercive promotion of the good would be less necessary if some optimistic assumptions about humans, for example, a strong natural desire doctrine, were true (3.1.2). If all humans wanted above all to develop their natures, there would be less need for publicity encouraging them to do so. There would also be less need for subsidies: Not only would the market cost of good activities be lower but also people would be willing to pay more for them. In chapter 3 we rejected the desire doctrine as unrealistically optimistic (3.1.4), and this has been reflected in our discussion of neutrality. Not believing that humans left on their own will always choose what is best, we have approved some state aid to help them do so.

At the same time, our position does not go to the opposite extreme and see humans as depraved. This was implicit in our composite case for liberty. That forbidding bad options prevents people's autonomously choosing against them would not be a relevant point if those who had the options could not resist them but would succumb willy-nilly to temptation. Similarly, the argument that coercion induces bad motives assumes that people who are not coerced can sometimes choose from good motives, an assumption I would affirm. In considering the tendency doctrines, we endorsed a moderate position like Green's: Humans have some tendency to the good and in favourable conditions can follow it, but they also have other desires that they can need help in resisting. It is therefore fitting that on a political question we also reach a position like Green's: generally favourable to liberty but rejecting the stronger ideal of state neutrality.

12

Equality: Abilities and Marginal Utility

From liberty we turn to the topic of equality and the worry that perfectionism favours unequal distributions of resources. Against a widespread view, I will argue that, especially given Aristotelian values, perfectionism's broad thrust is egalitarian, favouring substantial resources for all and not just for some elite. First, however, there is a more abstract topic.

12.1 Deep Equality

Some critics argue that perfectionism is anti-egalitarian, not just about the distribution of resources but more deeply. They say there is a fundamental moral requirement to treat all humans *as equals,* or with *equal respect and concern,* and that perfectionism violates this requirement. It thinks some people are better than others and therefore gives them greater standing.

12.1.1

One such argument concerns natural abilities. However perfectionism actually wants resources distributed, it is prepared in principle to give more to those with greater innate ability. This shows, critics argue, that it considers these people more valuable. Utilitarian and egalitarian theories care equally about all people's interests, but perfectionism gives the talented greater consideration.[1]

This argument cannot concern what is sometimes called "formal" equality. A distributively neutral perfectionism treats the equal achievements of all people equally, valuing a unit of excellence in one life no more or less than a comparable unit in a different life.[2] It does not value interests equally, if interests are defined by preferences or desires (2.2.3), but the reason is that it does not value preferences or desires. Why should it care equally about what for it is morally insignificant? Nor can the objection just be to the possibility of unequal treatment. Partisans of deep equality insist that it does not always require treating people the same. If some are handicapped or have special needs, we do not depart from equal concern but carry it through when we give such people special aid. Finally, the objection cannot just be that perfectionism thinks some lives are better than others. Utilitarian and many egalitarian theories think that happy, fulfilled lives are better and therefore to be sought for all. Where, then, is the special defect in perfectionism?

The objection about abilities trades on a false contrast. If a utilitarian or similar theory favours more resources for the handicapped, it may say these people are better to help or, with some straining, better as means to global utility or a just outcome. It will not say they are better in themselves. The same is true of perfectionism. It may say the gifted are better to help or better as means to overall excellence. That they count more in themselves, however, is simply false.

12.1.2

A second argument appeals to an ideal of philosophical neutrality. Because perfectionism has a substantive conception of the good, it is prepared in principle to give more to those who endorse this conception. Given their beliefs, these people are more likely to advance its values. But this, it is said, gives such people more than equal consideration and thereby treats them as better. To avoid partiality, political philosophy must be neutral about the good, not favouring any persons' ends above others'.[3]

To assess this argument, we must first disambiguate the phrase "conception of the good." In a broad sense, a person's "conception of the good" involves everything she desires, no matter what her reason for desiring it. In a narrow sense, it is defined by her judgements of intrinsic goodness. It contains only those things she desires because she believes they are good, either from her point of view or, as I will assume, agent-neutrally.

If political philosophy favoured some people's good in the broad sense, it would indeed violate deep equality, counting those people's desires more just because they were theirs. But the theory does not violate deep equality if it affirms a conception of the good in the narrow sense. Such a conception is impartial, applying equally to all humans. And its adherents will confirm this: If they imagine a possible world where they have false values and others have true ones, they will want the others' ends favoured. Their view of the good is therefore not tied essentially to them, and to favour it is therefore not to favour any particular people.

12.1.3

Other arguments confuse philosophical neutrality with the ideal of state neutrality discussed in the last chapter (11.4). State neutrality is an ideal for public policy: It is realized when government officials do not have as their reason for acting a substantive view about the good. Philosophical neutrality, by contrast, concerns the ultimate standards for judging policies, including a policy of state neutrality. It requires these ultimate standards to be neutral about the good.

Perfectionism does not exclude the possibility of state neutrality. Imagine that you are a perfectionist living in a society where everyone else is anti-perfectionist or has substantive values that are strongly opposed to your own. Here a state committed to your values may not be feasible; and of the political structures that are feasible, the best by your lights may be a neutral state. So you may campaign for a neutral state. Or you may believe that a perfectionist state, although feasible, would be counter-productive: Its attempts to promote perfection would more often work to hinder it. I do not believe that this last view is plausible; there are many ways, mostly non-coercive, that the state can effectively promote the good (11.4). Its possibility, however, shows again how facts that support state neutrality need not support philosophical neutrality.

Nonetheless, many anti-perfectionist arguments appeal only to such facts. They claim, for example, that questions about the good are controversial, or that in a pluralistic society no theory of the good can be the object of unforced agreement.[4] These claims, even if true, do not support any philosophical conclusions. They provide reasons, acceptable in principle to a perfectionist, why a perfectionist state may not be feasible or could not be instituted without great costs, especially for autonomy. I believe that in practice the claims should be resisted. There is substantial agreement about the good—more, I would say, than about the contentious theories of justice advocates of neutrality go on to defend. And disagreement counts for little against policies of non-coercive encouragement. Even if true, however, the claims do not show that our reasons for being impressed by them, or our ultimate political standards, should not be perfectionist.

Some arguments do try to establish the stronger conclusion, claiming that theories of the good are not just controversial but unknowable or meaningless.[5] But these claims are defensible only given a general skepticism about morality. Why should claims about the good be less knowable than claims about the right, especially when they *are* claims about the right, namely, about what it is right to desire in one's own life (5.1.2)? Other arguments claim that in political, as opposed to moral, philosophy we cannot go beyond what is accepted in our society. We cannot aim at true political principles, just at "willing political agreement."[6] But why should we accept this limitation? If what our society agrees to is true, why can we not seek a deeper explanation of its truth? If what we agree to is false, why can philosophical argument not show its falsity? Finally, some argue that philosophical neutrality reflects its own normative ideal, namely, a "Kantian" ideal of persons as free and equal.[7] Here we must ask: Does the argument claim that this ideal *is* universally accepted in our society? If so, we can both dispute the sociology and deny its relevance. If not, we can ask why political philosophy may use a controversial Kantian ideal but not a controversial Aristotelian ideal.

I will not say more about deep equality, for I do not believe it is a fruitful topic. The important egalitarian objections to perfectionism do not concern this highly abstract concept but rather the concrete issue of distribution.

12.2 Desert and Aggregation

Some perfectionists do advocate unequal distributions of resources. I have mentioned Plato, Aristotle, and Nietzsche, but as late as 1907 Rashdall could write:[8]

> All improvement in the social condition of the higher races postulates the exclusion of competition with the lower races. This means that, sooner or later, the lower well-being—it may be ultimately the very existence—of countless Chinamen or negroes must be sacrificed that a higher life may be possible for a much smaller number of white men.

Why this disturbing, even repellent elitism? Why has perfectionism sometimes been, as its critics charge, anti-egalitarian?

12.2.1

One kind of anti-egalitarianism must be set aside. Some philosophers think perfectionism favours unequal distributions because they think it uses a desert principle of the form "to each according to his x," with x replaced by "past perfection." This reading receives some support in the tradition: Aristotle, for example, says that "awards should be 'according to merit'" on every view, but according to excellence in particular on his.[9] But the reading ignores perfectionism's essential character. As a maximizing morality, perfectionism favours whatever distribution of resources most promotes aggregate perfection. This may sometimes mean giving more to those who have achieved more in the past, but the reason is not a backward-looking interest in desert—it is the desire for greater perfection in the future. Rashdall says that unequal rewards are "childishly unreasonable" if they mean that "every individual should be assigned sugar-plums in proportion to his moral or other merit," but are useful as means for promoting the good.[10] He defends unequal distribution as a device for increasing perfection, not for instantiating some intrinsically valued pattern.

12.2.2

A more serious basis for anti-egalitarianism is non-neutral aggregative principles. Some perfectionists, most notably Nietzsche, aggregate using maximax. They make society's goal not the sum or average of its members' perfection, but the greatest good of its most outstanding individuals (6.2). Maximax supports inequality on almost any assumptions about the world. If some people have more talent, they should be given more resources because only their perfections will matter morally. Even if talents are equal, there should be unequal distribution. Society should arbitrarily pick some of its members and devote the bulk of its resources to them, because this will produce higher heights than if wealth were shared equally.

In our earlier discussion, we rejected maximax partly because of its anti-egalitarian consequences. This rejection was by no means illegitimate: If broadly egalitarian intuitions can be used to criticize perfectionism, they can also be used to select among its variants. A concern for distributive equality, assuming we have it, can make us reject an aggregative principle that denies equality overtly, especially when, consequences apart, maximax does not fit our most serious perfectionist judgements, for example, about the importance of education for all (6.2.2).

Perfectionism could also be anti-egalitarian if it aggregated using the single-peak principle (6.3). When single-peak governs individual feats, there is value only in new achievements—new scientific discoveries or new climbs of mountains—and no value in repetitions. Society should therefore concentrate its resources on those who are capable of these achievements, and they are few in number. A different application of single-peak governs entire activities or forms of life. This too can have elitist consequences, requiring society to preserve an expensive perfection such as opera rather than make more of a cheaper perfection available to all.

Like maximax, however, single-peak is not acceptable in a pure narrow perfectionism. Not only can it have anti-egalitarian implications but also it does not fit our most serious judgements about, for example, educational policy (6.3.3).

12.2.3

As egalitarian perfectionists, Marx and Green both reject these non-neutral aggregative principles. Marx does not believe it is any justification of capitalism that its most favoured individuals lead challenging lives or that art and philosophy flourish somewhere within it. On the contrary, he condemns capitalism for stunting the development of most people. Green says that when others cannot achieve basic perfections, the pursuit of a higher good such as music is wrong:

> In some Italian principality of the last century, with its civil life crushed out and its moral energies debased, excellence in music could not be counted of actual and present value at all. Its value would be potential, in so far as the artist's work might survive to become an element in a nobler life elsewhere or at a later time. Under such conditions much occupation with music might imply indifference to claims of the human soul which must be satisfied in order to the achievement of a life in which the value of music could be actualised.[11]

Green's claim here is very strong: When others cannot achieve the basic perfections, pursuing music is not only wrong but also not intrinsically good. We cannot accept this moralistic claim, but we can accept the claim about right action that it presupposes. If music is not good, by Green's lights, in the circumstances he describes, it is because it is not morally right, and that is because the lesser perfection of others then makes greater demands.

Like Marx's and Green's perfectionism, our theory rejects both a desert principle of distribution and maximax and single-peak aggregations. But these features do not suffice to make it egalitarian. Even if society's goal is the greatest perfection of all its citizens, the best way to achieve this goal may be to distribute resources unequally. Some perfectionists believe this, and are anti-egalitarian not because of their theory's form but because of claims about its content.

12.3 Natural Abilities

12.3.1

What division of resources a distributively neutral perfectionism favours depends on three empirical issues: whether and how far people's natural abilities differ, whether resources are more useful for higher or for lower perfections, and whether perfection is competitive or co-operative (5.4.2).

Imagine that people's natural abilities differ greatly. Then some people can achieve more perfection given a fixed unit of resources, and to maximize perfection should be given more. If talents are similar, however, there is, if no argument for equality, at least not this argument against it. If resources contribute more effectively to higher perfections—if the marginal perfectionist utility of resources increases—there is again an argument for unequal distribution. Those who have already achieved the most, because they had the most before, can make the best use of further resources now. If the marginal utility of resources diminishes, however, the situation is reversed. Then resources are more important for lower achievements of perfection, and we should transfer resources to those who now have the least. The last issue is whether perfection is competitive or co-operative. If one person's perfection excludes that of others, it is not possible for everyone to achieve the good, and there is no point in aiming at this by distributing resources equally. If each person's perfection requires others', however, it requires resources for those others and therefore supports equality. The combination of claims most favourable to anti-egalitarianism is that natural abilities differ greatly, the marginal perfectionist utility of resources increases, and perfection is highly competitive. The egalitarian combination is that talents are similar, there is diminishing marginal utility, and perfection is co-operative.

In the next two chapters, I will defend the egalitarian claim about each of these issues, and show that perfectionism has egalitarian implications at least in actual conditions. (For the objection that perfectionism is anti-egalitarian in hypothetical conditions, see 13.2.2.)

12.3.2

Utilitarian discussions of economic distribution typically assume that people's utility functions—their capacities to derive satisfaction from resources—are roughly the same. But the parallel perfectionist assumption, that people's natural abilities are the same, is often denied. Plato and Aristotle think people's innate talents are very unequal, and Rashdall says, "nature has given to many Englishmen intellectual powers possessed by very few negroes."[12] If he defends anti-egalitarian racial policies, it is because of a belief about differing racial capacities.

Exactly how much talents differ is contentious, but the strong claims of Plato, Aristotle, and Rashdall are surely false. It is not true, as Aristotle believes, that women and "natural slaves" can live no better life than one of service to a husband or master,[13] nor are races other than the Caucasian incapable of culture. Although there are some differences in natural ability, they are nothing like as great as this.

Why, then, are the strong claims made? Some perfectionists may subconsciously use maximax and only talk about ability differences to justify their prior interest in the best individuals. But there is another possibility. In the social conditions known, say, to Aristotle, women and slaves could not do much to develop their innate capacities. They were not educated for public life and had little opportunity to exercise the highest rationality. Although it is fallacious, the inference from low achievement to low capacity is natural and may have tempted some perfectionists. It is still a danger today. Despite the greater equality of modern societies, children born to less favoured parents still have less opportunity. They grow up in less stimulating homes, attend inferior schools, and do not find around them the encouraging expectation that they will succeed at difficult tasks. There may be significant differences in people's present achievement of Aristotelian perfection; it does not follow that they all reflect differences in innate ability.

12.3.3

Let us agree that talents differ less radically than Aristotle and Rashdall believe. Can we move closer to a claim of equal abilities? Features of Aristotelian perfectionism suggest that we can. Because its account of quality is formal, this theory allows excellence to be achieved in many domains: in science and musical composition, but also in craft work, sports, and personal relations. Because it rests on structural properties of mental states, the good allows many kinds of content, and hence there is no one ability a human must have to lead a valuable life. Unless she is like Leonardo (7.4.5), her best life will concentrate moderately on one perfection, and her excellence will depend primarily on how some one talent is developed. But this talent can be any one among many—scientific, musical, athletic, or personal. This variety in turn suggests that people's overall abilities should be closer to equal. A person need not be skilled at science to achieve perfection, and the more alternatives she has to science, the more likely it is that she is suited to at least one. The more routes there are to perfection, the better chance each has to be able to travel by one. This claim does not deny that there are large differences in people's abilities in specific domains. Some are far better at science than others, and some have a greater capacity for personal understanding. But the differences between them should diminish when we consider overall abilities. If many different pursuits can have value, each person has many chances for some worthwhile potential. If we consider just their greatest talents, most people should have some worth developing.

If an inclusive conception of excellence equalizes abilities, a narrow one pulls them apart. This is important historically because many anti-egalitarian perfectionists have narrow conceptions of the good. Plato and Aristotle think serious perfection requires a capacity for philosophical reasoning, which is indeed unevenly distributed. Rashdall, too, attends specially to intellectual goods. To contrast these narrow conceptions with our own, consider productive labour. Aristotle thinks this labour is intrinsically valueless and counts as nothing the abilities of someone who can organize a manufacturing process or execute an intricate decorative design. He even thinks labour is a hindrance to perfection, so complete freedom from it is required for true excellence.[14] It is therefore no surprise that Aristotle recommends whatever organization of work is best for a leisured class of citizens and thinkers. But our theory is very different: It holds that productive labour can, if sufficiently varied and challenging, have positive value. In its own way, it can exercise sophisticated rationality, and it can be worth improving in the many at the expense of more refined perfections in the few.

In our modern societies, the dominant view of excellence is not as narrow as Aristotle's or Rashdall's, but it is still quite narrow. We find the greatest worth in verbal, mathematical, and managerial skills, but see less value in music, spatial discrimination, and emotional sensitivity. (There is acknowledgement for outstanding achievements in these fields, but lesser ones are valued almost at nothing.) This narrow focus further skews our beliefs about abilities. Not only do we ignore the effects of social inequality on present achievement, but we do not value all achievements equally. Looking only for a few talents, we undervalue those many people whose greatest abilities lie outside our favoured domains. This has a further effect. Given our societies' values, our schools and economies disproportionately encourage a narrow range of perfections. They promote logical and scientific reasoning far more than dance or visual sensitivity. (I say nothing of personal relations.) People with talents in these latter areas do not receive the same push to develop them or the same reward for exercising them. This lack of encouragement leads to a general cultural loss. If our societies tapped more of their artistic, expressive potential, they would be richer, more vital societies. And it further exaggerates our beliefs. Not only do we undervalue certain capacities, but our undervaluing prevents these capacities from being developed as far as others. People appear to us as generally unable who merely lack the abilities prized in our specific societies.

12.3.4

I have argued that, if there are many valuable talents, people's overall abilities should be more equal. This argument would fail, however, if the different talents all had a single innate ground. If there were one innate property that explained success in all fields, people with more of that property would be better at everything and people with less would be worse. That there is such a property is held by some psychologists, who call it general intelligence and say it is measured by IQ tests. Does their view re-establish unequal abilities?

Although the issue is controversial, many psychologists are abandoning belief in general intelligence. The statistical arguments for it are inconclusive,[15] and, as Howard Gardner argues,[16] there is mounting evidence that abilities to exercise rationality in different domains are independent. Different rational functions have been found to be localized in different parts of the brain. This physical separation of abilities suggests the possibility of developmental separation, which is confirmed by the existence of child prodigies and idiots savants. A prodigy may be brilliant at music but normal in other areas; an idiot savant may perform unheard-of mathematical calculations yet be handicapped elsewhere. These extreme cases merely dramatize what should be commonplace: that people with high ability in one area may have little ability in another. Someone who is brilliant at mathematics may be inept interpersonally; another's skill with spatial patterns may not help him see similar patterns in literature.[17] Finally, the integrity of the different abilities is confirmed by the fact that each has its own developmental history, through which both gifted and normal must pass, and by experimental results about tasks that transfer (or fail to transfer) across domains and forms of memory, attention, and perception that seem limited to one subject matter.

Using this evidence, Gardner develops a theory of seven distinct "intelligences": linguistic, musical, spatial, logico-mathematical, bodily-kinesthetic, and two personal intelligences (one concerning self-knowledge, the other the knowledge of others.) The exercise of all these intelligences involves formally similar operations, especially the grasping of extended contents and the manipulation of complex hierarchies. In Gardner's theory, however, ability in one domain does not imply a similar ability in others. That one can take in a chess position all at once does not mean one can "hear" a whole melody or "see" a whole proof. And this reinstates our argument for more equal abilities. If there are not just different perfections but independent abilities to achieve them, it is likely that most people have some talents and not others. Most people should be ineffective in some domains, yet have real potential in others.

We cannot adopt a view like Gardner's unreservedly; the existence of general intelligence is still hotly debated. Even if we assume this view, there clearly remain some differences in overall ability. Some people's greatest talent is so enormous that, even if it is moderately developed, they will lead better lives than most other people; others have no substantial abilities and even in ideal conditions can achieve just limited perfection. Even with multiple intelligences there remain some differences in overall talent, and this gives perfectionism a tendency to favour unequal distributions that is not found in utilitarian theories that assume identical utility functions (12.3.2). How far this tendency goes, however, depends on how much it is balanced by other egalitarian claims, of which the next is diminishing marginal perfectionist utility.

12.4 Diminishing Marginal Utility

Perfectionism cannot make the kind of diminishing marginal utility claim that generates a smooth curve on a graph, so the question is only whether it can say, roughly and in general, that resources are more effective in helping people move from low to medium levels of perfection than from there to the highest heights. If it can make this claim, the theory can say that unequal distributions give the rich what would do more good in the hands of the poor. Rawls thinks diminishing marginal utility is implausible in perfectionism,[18] but I do not see why. Especially given Aristotelian values, there are arguments for the claim both low on the scale of perfection, where resources are very important, and higher up, where they are not.

12.4.1

What does a person need for modest perfection—no great feats, but a reasonably informed, active life? First, he needs enough resources to survive. Then he needs whatever additional resources will make him feel secure. He will not pursue complex activities or make plans into the distant future if his day-to-day existence is in doubt. Beyond this, he needs education and the specific goods required for his specific perfectionist projects; these we will discuss later. But the first requirement for modest perfection is that minimum share of resources that makes for material security.

This requirement is clearest for leisure activities. No one can pursue research on colonial whaling or develop her violin technique who is uncertain of her evening meal: Science and art require freedom from material toil. But the point also applies to productive labour. Consider, for example, agriculture. Its most sophisticated forms are experimental: They discover the general principles governing plant growth in a locale by sampling different crops and techniques until the highest yields are reached. They have a long-term goal, namely, growing the most possible on some land, and work towards it by a series of connected innovations. But this experimental agriculture is not open to a peasant barely above starvation.[19] His traditional practices—growing rice, say, in a traditional way—may not be ideally suited to his land. It may be that if he switched to a cash crop such as jute he would do better. Even so, the alternative is not one he can discover. If he experiments with jute and his crop fails, he will have nothing to eat for the winter; if the jute price collapses, he will likewise have nothing to eat. His traditional ways assure him at least of subsistence, and that must be his first concern. In his precarious position, he must concentrate on the short-term and persist with whatever assures him of that.

This phenomenon is general. Sociologists describe a "culture of poverty" that develops in poor communities, both urban and rural. Its characteristic behaviours include never saving for the future, drifting between ill-paying jobs, and making purchases, even of food, only as the need arises, perhaps several times a day. Its psychological centre is a shortened time horizon: The very poor think mainly of the present, without extended or structured goals.[20] This constriction of aims harms the poor, creating a further obstacle to their escaping poverty, but it is also a natural response to their condition. To pursue long-term goals successfully, a person must be able to wait for results: She needs enough resources to see her through the possibly long period before her final aim is realized. The very poor lack these resources and therefore cannot make long-term gambles. If some elaborate project risks starvation in the short term, they would be foolish to attempt it, and they do not act foolishly.

The interpersonal analogue of present-centredness is self-centredness, which also characterizes the culture of poverty. Violence erupts quickly, with little thought for the harm to others, and personal relations are fragmented. Brothers who will fight to protect their sister's "honour" also swindle her to get money for drink.[21] But relations can only be fragmented among the desperately poor. Someone who is uncertain of eating cannot pause to develop nuanced feelings or pursue another's states as intrinsically valued parts of a larger goal. He cannot afford the short-term sacrifices involved in a long-term commitment. He must approach his relationships instrumentally, continuing them when they serve his present interests and abandoning them when they do not.

These claims should not be exaggerated. Sometimes the poor do pursue complex projects, and sometimes such efforts further their present aims. But these cases are of necessity exceptional. The material conditions of poverty make a certain habit of mind natural for the central issues in life, and this habit then tends to spread. Forced to take a shorter view of subsistence, the poor naturally take a shorter view of love, amusement, and violent anger. Nor is my claim that, once the poor have security, they will automatically develop sophisticated rationality. That would require a natural tendency doctrine of the kind we have expressly rejected (3.1.4). At any level of material resources, there are many other factors needed for excellence: energy, an accurate view of one's talents, a culture supportive of true values. No quantity of resources is sufficient for Aristotelian perfection, but a certain minimum is necessary. Someone who has security in the present can contemplate projects that stretch into the future; if he also has security for the future, his projects can concern something better than his material condition then. They can develop understanding or creativity and reach out to embrace other people. A person without security does not have this luxury. His main concern must always be his own present survival, which he must prefer to higher goods.

If material security is one condition for modest perfection, another is education. It is most obviously needed for theoretical perfection. Schooling gives us our basic knowledge of the world, as well as the tools, such as literacy and a grasp of scientific method, that we need to acquire more knowledge. At the same time, it can promote practical perfection. It provides the technical skills we need for many complex projects and develops mental habits of general utility. All education, however, costs money. If schooling in a society is private, only those whose families are reasonably well off can expect much from their lives. And for education to be public, the society as a whole must be reasonably well off. This provides some defence of the anti-egalitarianism of Plato and Aristotle. Given its limited productivity, their society could not afford schooling for all and had to choose between educating a few children and educating none. That it chose to educate a few was in the long-run best interests of human development, but it did mean a restricted existence for the many. When a society cannot educate some of its citizens, it denies them much chance at a valuable life.

12.4.2

For modest perfection, then, one needs enough resources for material security and a basic education. On its own, however, this fact does not amount to diminishing marginal utility. However important it is to move from zero resources to the minimum, it could be equally important to move from there to twice the minimum. But this is unlikely given the nature of Aristotelian perfection. The Aristotelian theory allows that perfection can be achieved in expensive activities. It can be achieved in building a baroque palace, voyaging through space, or managing a large investment portfolio. But these activities are not vastly superior to others that are far less expensive: Of the many routes to excellence, many are quite undemanding of resources. Consider the artistic and scholarly lives. They are often viewed as model lives, yet they hardly require great personal fortunes, nor do lives devoted to chess, athletics, or personal relations. When so much about Aristotelian perfection depends on one's inner states, and these can receive such different outward expressions, great wealth cannot be that important. Past a point it can increase the number of a person's options but will not add greatly to their quality. It may give her new opportunities for excellence but not significantly better opportunities. Of course, some people may enjoy the expensive options. They may get pleasure from building a baroque palace or managing large investments. But pleasure has no moral status in perfectionism and is no ground for unequal shares. What matters is how much wealth a person needs for *some* valuable life, and that amount is usually not much. In any field where she has talents, there are many excellent activities; because many of these are inexpensive, she can live well without enormous riches.

12.4.3

The classical perfectionists rarely discuss diminishing marginal utility as such, but many of them implicitly accept it. They have to, for they accept a stronger doctrine.

Many perfectionists claim that, past a point, additional wealth will decrease a person's excellence. It will corrupt her values, turning her away from true goods to the pursuit of material acquisition. Plato thinks a wealthy potter will only neglect his craft;[22] Aristotle warns that, just as one can have too little money, so one can have too much.[23] Even Rashdall says, "The scale of expenditure prevalent among the richest classes is as little conducive to their well-being as to that of the poorest."[24] These writers agree that, past a point, additional wealth is not important for increasing one's excellence. But they go further, holding that additional riches can be harmful, tempting a person from perfection into idleness and frivolity. For them, the marginal utility of resources not only diminishes but eventually becomes negative.

Given negative marginal utility, perfectionism would have an even stronger argument for equal distribution. It could say that great inequalities in wealth harm not only those who have less but those who have more. There would also be a related effect. Against arguments from diminishing marginal utility, there is standardly urged the need for incentives: Society must encourage the skilled to produce, to increase the total wealth available for distribution. Even with diminishing marginal utility, this retort has less force in more productive societies. If wealth matters less the more wealth one has, we should care less about economic growth the more our society is producing. But the point would be stronger given negative marginal utility. Then a society that used inequalities to boost production could become like Plato's "feverish" city:[25] Its attention shifted to expensive cars and clothing, it would embody less perfection than if its GNP were lower.

This last argument would have to be handled with care. Negative marginal utility is most plausible for privately owned resources, and they are not the only possibility. A wealthy society could manage many of its resources publicly and enjoy high GNP without large personal fortunes. In societies as rich as ours, some expensive perfections may now be justified, for example, particle research to increase scientific knowledge. Nothing requires that the facilities for these perfections be owned by their users. No physicist needs his own private accelerator, and there are positive arguments for public ownership: It is more efficient, because it provides access to more people, and it avoids the bad effects private inequality can have on power relations (12.4.4) and social attitudes (13.1.4). Public financing of expensive perfections does favour some people over others: Money spent on accelerators is not spent on other goods for other people. But the resulting inequality is less than if the same perfections were pursued using private wealth. Moreover, public ownership may avoid the damaging effects of negative marginal utility. If the greatest threat to values comes from large private holdings, public ownership may allow expensive perfections without corrupting anyone's values.

Negative marginal utility is a very strong hypothesis: It probably applies to some individuals and cultures, but not others. Even without it, however, a powerful egalitarian argument remains. If resources are more important for modest perfections than for outstanding ones, there is, on this basis, at least a defeasible perfectionist argument for equal distribution.

12.4.4

This argument for equality is strengthened if we consider instrumental goods other than material resources, especially freedom and power.

For modest perfection a person clearly needs power over herself. She must be free from others' control and able to form and pursue her own projects. But power beyond this, power over others, is less important. Like great wealth, it will increase the number of her options but not greatly affect their quality. This is the main reason why perfectionism opposes slavery and other extreme inequalities in power. Especially if autonomy is a perfection (11.1), slavery has terrible effects on the slave, denying him any self-direction or chance at valuable pursuits. And it really does little for the master. There is a special excellence to be found managing slaves, but it is not a singularly high excellence. Nor are the other excellences that slavery permits clearly better than those pursued independently of other people or, better still, in co-operation with them. (Consider South Africa under apartheid. Did its whites lead clearly better lives, by Aristotelian standards, than whites in other countries?) Given its rapidly diminishing utility, power should be distributed roughly equally.

This is a further reason why material resources should be distributed equally. To generate inequalities in power, one does not need the formal divisions of a slave or caste system; it is enough if some have much more wealth than others. Given the great influence of wealth, its possession will give them power over others, increasing the choices they can make and decreasing their fellows'. For the reasons just given, this is undesirable, and it can be so even if the inequality improves everyone's material condition. Even if the poor have more resources in an unequal society than in an equal one, they can be on balance worse off than if they had fewer resources but no one could dictate to them.[26]

12.4.5

The egalitarian tendency of diminishing marginal utility competes against the fact of (somewhat) differing abilities and the support this gives to (some) inequality. To reach a final perfectionist position, we must somehow balance these opposing claims.

In doing so, we must attend to the differences between instrumental goods. Some such goods are more subject to diminishing utility than others, and a person's benefit from some goods depends more on his natural abilities. To see this point, compare leisure time, or freedom from material toil, with money. (Leisure is not necessary for all perfections, but it is vital for many.) To some extent, these goods are connected. If some people are given leisure, others must supply their wants, and society loses whatever resources the former would create if they engaged in productive labour. But leisure and wealth are also partly independent and may call for different distributive policies.

As we have seen, wealth, especially private wealth, is strongly subject to diminishing marginal utility: Past a point, the perfections it permits are not significantly better perfections. The same is not true of leisure time. If Picasso has eight hours a day for painting, he may accomplish less in a ninth hour. But the parallel point does not hold for the more important judgements about entire lives. That Picasso has eight hours a day in his twenties is no reason to think he will accomplish less given eight hours in his forties or sixties. Indeed, if his art develops, he may then do his finest work. And this point seems to hold generally. Whereas the value of extra wealth in a life diminishes rapidly, that of extra leisure time does not.

A person's benefit from any instrumental good depends on his abilities, but again there are differences of degree. The utility of leisure time varies very much with talent. Indeed, to call someone specially able is just to say he can achieve more in a fixed time than others. Thus, Picasso is a more talented painter because he can create more of artistic value in an hour, day, or month. But this dependence is not so uniform for material goods. At the bottom of the income scale, the more talented do not use resources more efficiently: They do not need less money to survive or to feel secure. At the top of the scale, the dependence is limited by diminishing marginal utility. To express his talents a painter needs canvasses, brushes, and a studio, and a scholar needs books. But wealth beyond this level does not give them something their special talents can exploit.

These considerations, although rough, support a common conclusion: If perfectionism permits some inequalities, to reflect some differences in natural ability, they will be more of leisure time than of money. They will concern the opportunities people have, not the private fortunes they command. This recommendation is just what we find in the tradition. Plato's and Aristotle's most favoured individuals have more leisure than their fellows and a greater chance at the highest good, but they are not wealthier and do not live in significantly greater material comfort. Given negative marginal utility, this comfort could corrupt them, and, in any case, it does not suit their special gifts.

This result is significant because inequalities in leisure are less offensive intuitively than inequalities in wealth. Consider a policy of government arts grants. It gives selected playwrights and sculptors free time that others do not enjoy. If the grants are not large, however, and the artists not affluent, few find this objectionable. When the state finances a university system, it gives thousands of students and faculty the opportunity to pursue knowledge and discovery. Some academic research, especially applied research, may benefit all society, but many scholars, especially in the arts and sciences, study topics of purely intellectual interest. Even so, if their leisure is not accompanied by great wealth, few object. That professors and students have free time for inquiry, provided they use it that way, is not thought improper.

Why do inequalities in leisure seem less offensive? One reason is that they are less far-reaching. Unlike inequalities in wealth, they do not create further inequalities in freedom or power (12.4.4). Artists on grants cannot dominate others or control political appointments. Another reason, I believe, is that these inequalities serve genuine goods. Our intuitions recognize, even if subconsciously, that to give extra wealth is primarily to give the opportunity for more luxury amusement. If this amusement has no value, there is no justification for the inequalities it requires: These inequalities impose deprivation on the many for the sake of frivolity in the few. Inequalities in leisure, by contrast, promote perfection, a true good. They have genuine benefits to be weighed against their costs and thus are intuitively less objectionable.

13

Equality: Co-operation and the Market

The final egalitarian claim is that perfection is not entirely *competitive*. If it were always a necessary condition of one person's reaching a level of perfection that others do not, there could never be a universal advance in excellence, and there would be no point in pursuing such an advance by distributing resources equally. But is perfection in this way competitive? Or do the perfections of different people encourage each other?

13.1 Arguments from Co-operation

13.1.1

A claim of non-competition is more plausible for perfectionist than for utilitarian values (5.4.1). People often desire or get pleasure from things that cannot be shared, such as being sole owner of this or winner at that. But conflicts over scarce resources aside, why should one person's knowledge interfere with another's, or his developing skill prevent her doing the same? Aristotelian perfection consists largely in inner states that can be present in different people simultaneously; unlike satisfaction, it admits few conflicts other than ones over scarce resources. Even these conflicts are less serious, given diminishing (and perhaps negative) marginal utility. If additional resources matter only a little when one has a lot, even a moderately productive society can provide the material conditions of perfection to all its members. Some competition may remain over the means to more expensive (and slightly better) perfections or to avoid some socially necessary labour. Given just moderate prosperity, however, there is no bar to distributions that permit informed, achieving lives for all.

A stronger claim is that perfection is *co-operative,* so that increases for one are encouraged by increases in others and in turn make them more likely (5.4.2). This claim gives even stronger support to equality. To see this, consider the claim in an absolute form. If it were always impossible for one person to reach a level of perfection unless everyone else did so, there would be no point in any invidious inequality. Even if some people had greater innate abilities, they could not use these abilities unless everyone could achieve perfection alongside them.[1] Even with maximax, there would be no point in any invidious inequality. We could not create peaks above valleys or aim at anything other than a broad social advance in excellence.

Absolute co-operation is not plausible, and we can only consider a weaker claim. Yet, this claim may still have egalitarian implications. If there are ways in which the perfections of different people co-operate, there may be reasons why each person can live best in a society where the resources for perfection are available to all.

This idea is central to Marx's theory of distribution. Marx does not believe in innate ability differences, speaking instead of the "equal intellectual endowment" of all humans.[2] But his main positive argument for equality is co-operative: It is that in ideal social conditions "the free development of each is a condition of the free development of all."[3] According to Marx, each human can best develop his essential powers in a society where all develop their powers. Inequality in resources thus harms not only the poor—as follows from diminishing marginal utility—but also the rich. Even the development of the materially favoured is distorted by extreme inequality.[4]

This idea of Marx's—that each can develop best when all develop—is his greatest contribution to perfectionism. With specifically Aristotelian values, there are at least three grounds for the idea, or three co-operative arguments for equality.

13.1.2

The most obvious argument concerns the value of co-operation itself, or of collaborative action with others. This kind of action extends one's goals to include states of other people and, in face-to-face co-operation, adds dominance as well (10.2). Even a goal of no intrinsic worth has value when it is achieved in concert with other people, and goals that do have worth, such as knowledge and artistic creation, have their value enhanced. But co-operation has material conditions. To participate in joint action, a person needs security, education, and the specific goods required by her specific co-operative project. If this project's goal is an attractive neighbourhood, she needs a house and garden; if it is insight into Plato, she needs leisure and a copy of the *Dialogues*. The same holds for her partners. If she is to achieve the good of co-operation, others must be able to achieve it, and this requires that others have the same resources as she has.

This first argument has an obvious limitation. Because no one co-operates with all other persons, no one strictly requires resources for every other citizen. But much that is persuasive remains. If co-operation is a good, it will be most available to each person when all persons are potential collaborators with her, and the opportunities for joint action are in that sense most extended. This is especially so given the many kinds of co-operative venture that are possible. If there are many potentially worthwhile joint endeavours, some more suited to a person's talents than others, she will want to find people with co-operative interests close to her own, which will be easiest when everyone has the means to attempt co-operation. There is also a subtler point. One condition for successful co-operation is experience in co-operating, and someone joining a group will want its other members to have this experience. They can have acquired it in earlier joint projects with her, but they may contribute more if their past activity was with others, which requires those others to have had resources. Even if no one co-operates with everyone, an ideal society may contain a series of overlapping co-operative groups. Within each group, people benefit directly from the activities of the other members, but they also benefit indirectly from events in other groups that enhance their present partners' co-operative skills. This second benefit can be iterated indefinitely. Because co-operation is improved by (other) co-operation, each citizen may do best when every citizen, including many she never meets, has the material resources co-operative action requires.

An extension of this first argument concerns personal relations. These flourish when two people can attend accurately to each other's needs and pursue goals that unite them in a single intended state (10.3). But here, too, there are material conditions. No one can develop nuanced sensitivity or abandon self-concern who must always be worrying about his physical survival (12.4.1). For one person's relations to flourish, therefore, others need minimum resources. Of course, each person's closest relations are confined to a few people, but he has thousands of less intimate contacts, and it is desirable if in all these contacts there can be some reciprocated sensitivity. If fruit vendors and bank tellers can deal attentively with him, he can deal attentively with them. Again, there is a point about overlapping groups. My closest friends will have other closest friends, who have other closest friends, and so on. For my relations with my partners to go best, those partners need interpersonal experience, which may require security and education for people I never know.

13.1.3

A second argument turns in a different way on personal relations. For these relations to flourish, people must be open to each other's characters and confident about expressing their own. But economic inequality, especially when it has hardened into a system of economic classes, can inhibit these traits. Interaction across a class barrier can become strained and unnatural: Both the rich and the poor can feel ill at ease with someone from utterly different circumstances. Even worse, there can develop arrogance on the one side and servility or resentment on the other. The rich, overly impressed by their material advantages, can condescend to those who are less fortunate; the poor can suppress themselves in cross-class encounters or, if they do not, develop a blinding resentment. Nuanced human interaction is an important good to be sought for all. Economic inequality, by creating large differences between people, makes it harder to achieve. Large income gaps direct attention to what separates people and so poison their sense of community.

Again, it will be said, this argument has limitations. Whatever happens across classes, nuanced relations are possible within them, between rich and rich or poor and poor. But each citizen has interactions with thousands of other citizens, and it is desirable if all her contacts can express some attentive concern. More importantly, it is doubtful whether attitudes can be hived off in the simple way the objection assumes. If a person is self-effacing or resentful with those who are richer, may this not affect her behaviour with her material equals? If someone grows used to ignoring the poor, may she not come to ignore everyone? Personal relations do not divide up neatly. Attitudes developed within one context can spread beyond it, especially attitudes born of economic inequality. Class division encourages the general thought that people come with places in a hierarchy and are to be treated differently as their standing is different. This thought can colour all one's dealings with others and affect even one's most intimate relationships.

13.1.4

The final co-operative argument concerns another indirect effect of inequality. Because economic differences are so conspicuous, they draw attention to their subject matter: material possessions. They direct ambitions towards the accumulation of wealth and away from higher values. A society of material equals is unlikely to be one in which people's main aims are monetary, but a stratified society, by allowing gaps in income, makes economic standing salient and thus can distort people's values. When material possessions do so much to distinguish people, they naturally become a prime object of desire.

This argument differs from the earlier one about negative marginal utility (12.4.3). There it was a person's absolute level of wealth that could distort his values; here it is large differences in relative wealth. At any level, the present argument says, large inequalities in possessions make possessions salient and so direct attention to them.

The effects of material interest need not all be bad. Many money-making activities, including business activities, demand sophisticated rationality and can therefore, given formal measures of quality, embody perfection. More often than not, however, the methods that earn the most do not most stretch human capacities: Their "efficiency" is at the expense of challenge and complexity. There are also more subtle effects. To have the most value, work must be chosen not just instrumentally but in part for its own sake (9.5.4): A worker may aim to make money, but he should also will his tasks for themselves. But this double intending is less likely when one's whole culture stresses material acquisition: Then the temptation is to value work only as a means. There may also be consequences for the use of leisure. If people's worklives aim so centrally at money, may their leisure choices not suffer? May they not prefer material consumption to extending their knowledge or pursuing difficult goals? Our formal account of quality allows many activities to have value, including money-making activities. But there are still degrees of better and worse, and a commitment to correct values is needed for any pursuit to have its greatest worth. If material inequality undermines this commitment, it does perfectionist harm.

Some high-minded perfectionists envisage a society where all see others as partners and there is no competition among them. Perhaps this ideal is achievable; perhaps it is not. Even if it is not, however, it matters how humans express their (on this view, ineliminable) competitiveness. In our societies people compete primarily for material goods, which have no intrinsic worth. Without inequality—and without the false values it engenders—they could compete instead in excellence. They could strive to outdo each other in knowledge, discovery, or creative expression. Then their competitiveness, instead of hindering perfection, could spur them on. This is a further argument for distributive equality: By encouraging human competitiveness to aim at true goods, it makes an unattractive trait serve valuable ends.

Would competition in excellence not fall afoul of the second co-operative argument? Would it not lead people to view others hierarchically and thereby undermine personal relations (13.1.3)? If it would, there are limits in human psychology to the full attainment of excellence. Still, competition in achievement seems less of a threat to personal relations than competition for material goods. The objects of material striving—wealth, status, possessions—are usually valued on a relative scale. One cares not about the absolute level of one's income, but about whether it is more or less than others'. Perfection, too, can be valued on a relative scale; a painter or historian can care that she be the best of her generation or better than a particular rival. But here it is easier to value one's work also on an absolute scale, to see it as good in itself or as fulfilling some human potential. Where material goods are commonly valued just relatively, with perfection one naturally also takes an absolute view. And the more this view is present, the less personal relations suffer. Someone who sees her own work as good in itself must see others' work as good; given this attitude, she must see others in part as deserving respect.

There is also a difference in the complexity of the two valuings. The scale of material standing is essentially unidimensional: Some people simply have more and others less. But the scale of Aristotelian perfection combines many different values. I may be more accomplished than another person in my chosen specialty and even more accomplished, all things considered. Given the diversity of Aristotelian perfections, however, there are bound to be some domains where he outdoes me, and achieves not only more than I do but more than I could. If I accept Aristotelian values, I must acknowledge this fact. I must acknowledge that almost everyone is more accomplished than I in some pursuit I consider valuable (a great liberating effect of adopting Aristotelian values). This acknowledgement makes condescension more difficult and encourages appreciation for all.

There are, then, two related effects. Not only can material inequality distort people's values by fixing their attention on material possessions, but it also channels their competitiveness where it most threatens personal relations. With distributive equality, they might pursue excellence for its own sake and compete, if doing so is inevitable, in ways that had some good effects, and fewer bad ones.

13.2 Illustrations and Limitations

As fully assembled, the perfectionist case for equalities claims that, although there are some differences in natural abilities, most people have some talents worth developing; that there is diminishing (and perhaps negative) marginal perfectionist utility; and that, for the reasons just given, each can best develop his humanity in conditions where all can develop their humanity. All three of these claims are made by the egalitarian writers influenced by Green, and especially by the best known among them, L. T. Hobhouse.[5]

13.2.1

Hobhouse's treatment of distribution starts from something very close to an Aristotelian account of excellence. What gives a life value, he claims, is the "harmony" of a person's intentions or the degree to which they form a "system" and give her a "personality."[6] This account of perfection supports, first, Green's case for liberty (11.2). Because "human personality is that within which lives and grows, which can be destroyed but cannot be made,"[7] the state must leave its citizens free to achieve their own excellence. It also supports egalitarian distribution. In *Liberalism*, Hobhouse defends a "Liberal Socialist" programme including steeply progressive taxation, the nationalization of all privately held land, and the provision by government of a guaranteed minimum or "living wage" to all.[8]

Hobhouse allows that there are some differences in innate ability and that some incentives are needed to encourage the able to produce. But the annual incentives need not be large—he envisages 100% taxation on annual incomes over £5,000 (in 1911 currency)—nor are the ability differences so great as to exclude the mass of people from excellence. On the contrary, "a fulfilment or full development of personality is practically possible not for one man only but for all members of a community."[9] More positively, Hobhouse's main arguments for equality stress the vital need, for any perfection, of enough resources for security and independence. The idea of a "living wage" is the idea of "such a wage as would provide the real minimum requirements of a civilized life" or "the means of such a healthy and independent existence as should be the birthright of every citizen."[10] More generally,

> Needs differ in urgency. A certain minimum of food, clothing, etc. may be regarded as of absolute necessity. Certain additions to these add greatly to comfort and efficiency. Further additions have less effect. A law of diminishing returns applies pretty rigorously to the relations between healthy development and physical conditions. It is clear that the most urgent need is invariably to be preferred. Thus the minimum necessary to physical health and the normal growth of faculty takes precedence of all other personal claims, and in general so far as they are distinguishable and classifiable necessaries take precedence of comforts, comforts of luxuries.[11]

Finally, Hobhouse uses arguments from co-operation. For him as for Green, there exists a "common good" (5.4.2), one involving not just no conflict among different people's perfection but positive mutual support. Each person has "possibilities of development such as not merely to permit but actively to further the development of others."[12] If the rich use their wealth to educate the poor, they will increase their own perfection, and they can expect further increases when the poor, having developed, can join in new collaborations with them.

The last argument is developed further by a later writer in this tradition, R. H. Tawney. Tawney's principal moral value is community, or harmony among the members of a society. But community requires material equality because great gaps in wealth produce not a common culture, but "servility or resentment, on the one hand, and patronage or arrogance, on the other."[13] To avoid these distortions, institutions "should be planned, so far as possible, to emphasize and strengthen, not the class differences which divide, but the common humanity which unites" people.[14] This is our second co-operative argument, about personal relations. Tawney also cites the distortion of values. A society is civilized, he says, when its conduct is guided by "a just appreciation of spiritual ends." Great contrasts in wealth and power "do not promote the attainment of such ends, but thwart it."[15] They create a commercial or plutocratic spirit, whose tendency

> is to pervert the sense of values. It is to cause men, in the language of the Old Testament, "to go a-whoring after strange gods," which means, in the circumstances of today, staring upwards, eyes goggling and mouths agape, at the antics of a third-rate Elysium, and tormenting their unhappy souls, or what, in such conditions, is left of them, with the hope of wriggling in. It is to hold up to public admiration sham criteria of excellence.[16]

When some are rich and famous, their shallow lifestyles become an object of envy for all.

13.2.2

The composite perfectionist case for equality does have limitations. If some people have greater innate abilities, they should sometimes be given greater resources. Even when overall talents are the same, some people have their greatest ability in more expensive areas. In conditions of scarcity, it may be right for them to receive fewer resources, but given material abundance they should have more. If society can afford everyone's perfection, it should give them the extra resources they need to realize their best potential. Finally, larger shares for some can be justified by the benefits this distribution provides to others. These benefits may be economic, as when some receive incentives to greater productivity, or they may directly concern perfection. If actors and ballet dancers have extra resources, they can offer finer aesthetic appreciations to their audiences; if researchers have accelerators, they can increase knowledge in future generations. Within the constraints of diminishing marginal utility and the arguments from co-operation, there can sometimes be net benefits from unequal distributions.

In discussing ability differences, I argued that the inequalities they justify are more of leisure time or opportunity than of private wealth, and that these are less offensive inequalities. They are less prone to create further inequalities in power and they promote perfection, a true good (12.4.5). Similar points apply to the other perfectionist grounds for inequality. Imagine that in a prosperous society everyone has a reasonable chance at excellence, but those with more expensive talents are given more resources. If they achieve the same overall excellence as others, how is the inequality objectionable? Or imagine that the greater perfection of some indirectly benefits others, for example, that researchers with laboratories increase everyone's knowledge. If inequality has this result, is it not desirable?

These points may also answer a deeper objection to perfectionism, namely, that its egalitarianism depends on contingent empirical facts. We can imagine possible worlds where people's abilities differ greatly or perfection is highly competitive, and in these worlds perfectionism favours very unequal shares. Does this not show that the theory is unacceptable? This kind of objection is often thought to tell strongly against utilitarianism. If wealth did not add diminishingly to people's satisfaction or if their utility functions differed greatly, utilitarianism would not favour equality, and this is often thought to warrant its rejection. But the issue is less clear with perfectionism. Whatever some philosophers have claimed, our intuitions recognize that mere pleasure is not a serious value. So when we imagine beings who are much more efficient at getting pleasure from resources, we think utilitarianism is wrong to say they should have more: That would deprive others to promote something essentially trivial. Perfection is not trivial, however, and our judgements about the parallel perfectionist cases are far less certain. If there were beings with vastly greater abilities than ours, who could know, create, and achieve vastly more given resources, would we begrudge them a larger share? If their gains would outweigh all ours added together, might we not acknowledge their greater claim? I have not argued that perfectionism is the one true morality; it may need to be supplemented by other principles, including egalitarian principles. But the supplementation is less obviously needed than in rival theories such as utilitarianism. Precisely because the inequalities perfectionism approves advance excellence, a true good, they have serious benefits to be weighed against their costs and are therefore morally less objectionable. Our intuitions about distribution have developed in societies where the most common effect of inequality is to increase consumption and idle amusement for the few. Because this effect is morally trivial, we are right to condemn its cause. For the same reason, however, it is less clear that our repugnance transfers to inequalities that promote excellence. The issue is difficult, but in the possible worlds where perfectionism permits great inequality, it may permit what is there morally appropriate.

13.3 Property and Property-Freedom

To this point we have taken certain distributions as given and inquired about their effects: We have assumed that some mechanism produces a division of resources and asked whether the most good results when this division is equal. But we have ignored the process of distribution and any further effects resulting from that. For a complete discussion, we must consider the consequences not only of different distributions but also of the different mechanisms that can bring them about.

The best-known distributive mechanism is the *free market,* where goods are privately owned and circulate by the owners' free choices. The outcome of market distribution depends on circumstances and owners' desires, but it is rarely strict equality. As a result, egalitarianism requires some interference with the free market. Some perfectionists oppose this interference, arguing that the maximization of perfection requires pure market distribution. They have two weaker arguments for this claim and one stronger one.

13.3.1

The first argument claims that perfectionism supports a derivative right of private property, or a right to own and transfer possessions, and that this right is violated by interference with the market.

The argument originates in Hegel's account of property as the "embodiment of personality"[17] and is stated most fully by Green and Bosanquet. For them "personality'' means, in particular, acting on extended, dominant ends. Green contrasts true ''appropriation,'' which ''implies the conception of himself on the part of the appropriator as a permanent subject for whose use, as instruments of satisfaction and expression, he takes and fashions certain external things,'' with the mere passing use of objects to satisfy ''this or that want.''[18] Bosanquet rests the right of private property on the idea that ''the inward or moral life cannot be a unity unless the outward life—the dealing with things—is also a unity.''[19] In other words, property is justified as a means to the good life, where that involves the achievement of sophisticated ends. Property encourages this achievement partly by giving people the security they need to attempt long-term projects (12.4.1), but it also supplies the material for long-term projects. On permanent possessions a person can impose the various stages of her lifeplan, and in these possessions she can later see the marks of her past activity. Her property is a means both for affecting her future and for connecting with her past; in both ways it contributes to a unified life. Bosanquet sometimes puts this point aesthetically. A person who owns a house can express in its arrangement and decoration a (perhaps developing) sense of style and, having done so, can then live surrounded by the products of her past creativity.[20]

Although impressive, this justification of property does not support anti-egalitarian conclusions. It is consequentialist, citing the benefits to an owner of having possessions. But it must also consider the costs to non-owners, particularly given the fact of diminishing marginal utility. Modest possessions may extend someone's goals at little cost to others. Vast wealth, however—a palace, factory, or bank—brings small gain to the owner and much greater loss to others. Counting the effects on everyone's perfection does justify property rights, but not unqualified property rights. These rights entitle people to own something, but not to amass huge personal fortunes.

13.3.2

The second argument extends Green's case for personal liberty (11.2). If perfection requires freedom, the argument claims, it requires the freedom to use and transfer property, which in turn requires a free market. Just as it is wrong to interfere with people's sexual activities or choice of career, so it is wrong to interfere with their free decisions about the objects they own.[21]

Like the previous argument about property rights, this one has an undeniable first premiss. Property is of little benefit to an owner unless he has some control over its disposition, and there is therefore a presumption in favour of free use. But we must again consider the effects on non-owners as well as owners, and when we do, we find that some freedoms concerning property are more worth granting than others. If someone owns a house, it is clearly important that he be able to decide what happens inside it while he is alive. It is less important that he be able to bequeath the house after he dies; the perfection he can gain through bequest is clearly less than what others can lose through unequal distribution. For the same reason, it is less important that he have the other freedoms to transfer that constitute the free market.

Imagine that, given egalitarian redistribution, someone who earns $50,000 in a year must pay one fifth of it in taxes. Those who defend the market in terms of freedom cannot claim that this deprives the person of an important freedom to use $50,000, for not even the pure market guarantees him $50,000. If his employer valued his work less highly or if more people shared his skills, even pure laissez-faire might give him just $40,000. The market already embodies a thoroughgoing dependence on other people, so redistribution cannot be said to restrict a freedom employees would be sure to have without it. If anything, redistribution restricts the freedom of *employers*. In the pure market, someone who has money to spend on wages can see all of it transferred to her workers: If she has $50,000, she can see that her workers receive $50,000. Given redistribution, this is not so; she can transfer only $40,000 to her workers and in that respect has her options restricted. But how important is this? How vital is it to her perfection to be able to spend all her money hiring others rather than just a large portion? As with free bequest, the small loss to an owner from redistribution is not sufficient to outweigh the large benefits to others.

This argument explains why, to return to an earlier topic (11.4.1), it is no serious objection to perfectionism that it may tax citizens to fund perfectionist activities such as science and the arts. This funding violates state neutrality, but that is not a plausible ideal (11.4.1), and the funding does not violate any reasonable ideal of liberty. Classical liberalism protects those choices through which people can shape their own lives and determine their major goals, but offering others money is not central for these purposes. If advancing the good requires limiting people's ability to bequeath possessions or hire at high wages, it does not infringe a freedom seriously worth protecting.

13.4 Self-Reliance versus Dependence

The first two arguments for the market concern the effects of redistribution on the wealthy people who must finance it. The third and more serious argument concerns its effects on those who receive redistributed wealth. It is the main argument used by Bosanquet, the most important perfectionist defender of the free market.

13.4.1

Because perfection is active, Bosanquet argues, it is best achieved when people are self-reliant and find material well-being through their own work and initiative. They take thought for their future and themselves do what will make it secure. The free market encourages self-reliance by making people responsible for their own fate: if they do not work, they suffer; if they fail to protect against illness or old age, they will pay a price later. These incentives, however, are undermined by egalitarian redistribution. A guaranteed income frees people from the need to be active, and state pensions make unnecessary that prudential foresight that binds a life into a unity. Where free competition encourages rational self-direction, redistribution makes for dependent lives that never look beyond the present.

Bosanquet's is not the cynical argument that the poor should be self-reliant to be less of a burden on the rich. It is a perfectionist argument that appeals to genuine goods in the poor's own lives. It is currently enjoying a revival, as conservatives on both sides of the Atlantic rediscover "Victorian values" such as thrift and independence. These conservatives' objection to the modern welfare state is not just that it is costly, but also that by creating long-term dependence it encourages undesirable traits in the poor.

This is a serious argument, one with which an egalitarian perfectionism must make its account. If it aims to defend pure laissez-faire, however, as against any redistribution whatever, it cannot succeed. In the strong form needed to justify the pure market, the argument's premises are both individually dubious and in considerable tension with each other.

13.4.2

First, the argument is implausibly optimistic about the effects of market distribution. It assumes that all people who are able and willing to work in a free market economy can earn enough to secure their future, support a family, and perhaps decorate a house. This assumption is naive. During the Great Depression, many thousands of men and women crossed Canada and the United States looking for work, but none was available. Even growing economies, economists now tell us, have a level of "structural unemployment," a percentage of the workforce that cannot be employed even in ideal conditions. How does the market care for these workers? In a prosperous time it may be true of each person that, if she looked harder for work, she would find it. But her success would depend on others' not looking harder, and it would not be true that if all looked, all would work. The free market always leaves some people unemployed, and in bad times it leaves many people unemployed. Without redistribution, what gives these people a chance at a valuable life?

The argument is also naive about the stability of the market. Bosanquet values laissez-faire for attaching a constant schedule of rewards and punishments to people's actions and thereby encouraging long-term projects; hence its superiority to his prime bogey, the chance fluctuations of unorganized charity. In a market economy, however, a worker may develop a skill and find, ten or twenty years later, that technological change has made that skill obsolete, or a town may have to shut down because management of the local enterprise has been incompetent. To the miners of South Wales or the steelworkers of Pennsylvania, the market is hardly a mechanism that consistently rewards foresight and self-reliance. On the contrary, it is a system of arbitrary and unpredictable forces, where carefully thought out plans can be wrecked by events occurring many miles away.

Nor is it only the economy about which the argument is optimistic. Bosanquet assumes that when the market gives people responsibility, they will accept it. If they are left to save for old age, they will see the costs of imprudence and avoid it; if they have to feed their children, they will. What if this does not happen? What if market incentives do not work and some people still pursue only present pleasures? Then even the free market will produce people who have failed to save and are now indigent, and it will produce children who are too ill-nourished to learn. It will be too late to administer bracing incentives, and society will have to choose between offering assistance and letting the families of the imprudent go under.

More specifically, Bosanquet is naively optimistic about the poor, thinking their material condition no bar to their acceptance of responsibility. He is entirely candid about this assumption. Against the view that social conditions affect people's conduct, he maintains that "character is the condition of conditions"[22] and that "our growing experience of all social 'classes' proves the essentials of happiness and character to be the same throughout the social whole."[23] Taken strictly, these claims eliminate another argument for the free market, namely, that it encourages economic growth. If all the excellence that matters can be achieved on the wages of an honest factory hand, what need is there for great engines of productivity or a rich managerial class? Even apart from this point, however, the assumption is false. The higher perfections—scientific research, art, politics—require resources far beyond those of an ordinary worker, and even the more modest perfection Bosanquet talks of has material requirements. As we have seen (12.4.1), extreme poverty encourages habits of mind that are inimical to excellence. It forces people to think about their survival in the short term, and the resulting present-centredness then tends to spread: It creates a "culture of poverty" that inhibits perfection in many domains. This point was appreciated by Bosanquet's contemporaries. As one of them said, the poor may be "thoughtless and extravagant," but the conditions under which they live "themselves furnish an education in improvidence."[24] To seek the long-term rewards offered by the free market, one must be able to think of the future, which requires at least material security in the present.

If Bosanquet is optimistic about the market, he is extremely pessimistic about the effects of redistribution. He assumes that the poor, or many among them, will respond to state aid by abandoning thrift and lolling in the pleasures of the present. Why must this follow? Why may people freed from insecurity not turn to other goods such as personal relations or the exercise of skill? Why may they not even turn to higher goods? Buying health insurance or an annuity is hardly the highest activity possible for a human. So why may state assistance not release people for better goals and richer forms of rationality? Bosanquet is highly optimistic about people's response to the market—why not extend this optimism to the effects of equalization?

These questions are especially pressing because, for all his dire claims about the effect of unearned income on the poor, Bosanquet makes no similar claims about the rich. He does not call for restrictions on large gifts, bequests, or windfall profits. If it is corrupting to receive a welfare cheque, however, should it not also be corrupting to receive an inheritance or dividend cheque? If the evil is discouragement from activity, should it not be greatest when what is received makes activity unnecessary for one's whole life? In a free market economy, many people receive what they have done nothing to earn and live on it for the rest of their lives. Yet Bosanquet does not find this objectionable. He might claim that the difference rests on innate abilities: Whereas the poor are capable only of managing their finances and have nothing when that is taken away, the rich can use leisure for higher achievements in the arts and sciences. This claim is odious, however, and, not surprisingly, is never explicit in perfectionist arguments for the market.

13.4.3

To justify the pure market, then, Bosanquet needs empirical assumptions that are not only dubious but also in considerable tension with each other. And this is not his argument's only flaw. Its concern is how economic arrangements affect people's attitudes, yet in this area laissez-faire itself is suspect. As a distributive mechanism, the pure market has several regrettable psychological effects.

First, the market's emphasis on economic competition can foster a damaging egoism. In a market economy, businesspeople strive to outsell their rivals and even, at times, to drive them out of business. In private life, people are encouraged to outconsume and outpossess. When a system so consistently pits people against each other, how can relations between them not suffer? How can the competitiveness of their worklives not colour their attitudes generally? Earlier we saw how inequality in distribution can corrupt personal relations (13.1.3); here a distributive mechanism, by encouraging and rewarding competitiveness, can have the same effect.

Second, the market promotes false values. Caught in the web of competition, each firm in a market economy must constantly strive to increase its sales. If it does not, it will be overtaken by its competitors and lose its market share. Therefore, each firm must promote material consumption. It must advertise the charms of buying as against other uses of resources and so propagate consumerist values. It is true that the market increases productivity, which can improve people's lives. As G. A. Cohen notes, however, gains in productivity can be used in two ways: to increase production or to relieve workers' toil, by giving them more leisure or more challenging work. In undeveloped economies the first use is best, but as wealth increases, the second becomes preferable. Yet the free market cannot make this choice. Forced to compete with each other, its firms must seek greater sales and in so doing promote consumption at the expense of higher goods.[25]

This aspect of the market means that, contrary to Bosanquet's picture, laissez-faire sends people very mixed messages. On the side of production it needs workers who are hardworking and disciplined, and its labour market rewards those who display these traits. On the side of consumption, however, it socializes workers to buy and buy now. It progagates an ethic of immediate self-gratification, of acquiring new amusements as quickly as possible.[26] In one of its aspects, the free market does reward thrift and responsibility; in another, it encourages an antithetical self-indulgence.

13.4.4

I have considered this third argument as defending pure market distribution and found it wanting, but it may be more persuasive in supporting a weaker conclusion. Hobhouse and the writers of his school were social democrats rather than socialists in part because they saw some justification for property rights and property freedom. But they also believed that, once people have the essential minimum of resources, they need incentives to further valuable action. Bosanquet's views did not fit the destitute, but for the materially secure his goads to independence were appropriate. If we follow Hobhouse, we may let the third market argument qualify our egalitarianism. We may limit the quantity of redistribution we favour, to leave some room for the market, or we may prefer some redistributive schemes to others. If full employment policies, for example, do less to undermine individual initiative than do welfare payments, we may on that ground prefer them.

There are difficult issues here that are rarely faced by egalitarian perfectionists, especially Marxist perfectionists. Asked about the future, Marxists typically reply that in conditions of equality people will have work and social co-operation as their "prime wants": They will not need incentives to do what they already most desire (3.1.2). There is no evidence for these psychological claims, however, and our long human experience against them. Even in areas far removed from the market, we see our easy tendencies to laziness, competitiveness, and greed. If we take humans as they are and always have been, we find that they often need external incentives to realize their best potentials, and this raises the question of whether the society most favourable to perfection may not use some stimuli from the market.

I have no settled answer to this question. More specifically, I have no settled view of whether the economy most likely to promote the perfection of all is a market economy with the redistributive elements of the welfare state or some non-market, socialist alternative. Of the former we must ask: Does its reliance on the market necessarily imply competitiveness and the spread of false values? Can education and moral reflection by individuals counteract the systemic distortions of a competitive economy? Of the latter we must ask: Does it fit humans as they are, or is it only workable given implausibly rosy assumptions about their present or potential motivation?

The empirical considerations that would answer these questions lie beyond the scope of this book, but we can still draw a general conclusion. Many critics claim that perfectionism applied to issues about distribution supports an elitist programme like that of Plato, Nietzsche, or Rashdall. I have argued otherwise. The most plausible perfectionism aggregates values in a distributively neutral way, and its Aristotelian values make three empirical claims plausible: that most people have some significant abilities; that resources are more important for lower than for higher perfections, so there is diminishing marginal perfectionist utility; and that, in several ways, each person is best able to develop in conditions where all can develop. Together these claims give perfectionism a strong but defeasible tendency to favour material equality. They support at least social democracy and may support a more radical egalitarianism.

Think again of our modern Western societies. Their more prosperous members are doing fine with their university studies, their interior decorating, and their business careers. What cries out for attention is the plight of those millions of people, both within the industrialized world and outside it, whose material condition prevents them from acquiring any organized knowledge or achieving any truly stretching goals. While others have more than they need for valuable activities, these poor are denied any substantial exercise of their essential powers. This is what Aristotelian perfectionism finds most appalling and what it most wants to see altered.

14

Conclusion

This book has described perfectionism, in its most plausible Aristotelian version, from the theory's initial ideas about human nature to specific claims about intrinsic values and political practice. The description has revealed many attractive features of pure perfectionism, but it has also shown that there are moral ideas the theory cannot capture. Pure perfectionism does not find intrinsic value in pleasure, not even pleasure in what is good, nor does it find intrinsic disvalue in pain. It does not view moral virtue as intrinsically preferable to moral vice; considered apart from their effects, vicious acts can embody high perfection. Because the theory cannot use formally egalitarian aggregative principles, it favours equal distributions of resources only in certain contingent circumstances. It does not guarantee individual rights, such as a right not to be killed, or place other constraints on the pursuit of good consequences.

Readers troubled by these omissions may wish to abandon pure perfectionism for a pluralist theory in which perfectionist ideas are weighed against others concerning pleasure, equality, or rights. This has always been a background possibility: to include perfectionism in a more inclusive moral view. But if it counts against pure perfectionism that it does not capture certain moral ideas, it also counts against those ideas that they cannot be derived from an ideal of human nature. The narrow perfectionist idea—that the human good consists in the development of human nature—is not only attractive but also makes a peremptory claim. It is of sufficient depth and power to present itself as not one moral idea among others but the foundation for all morality, and it therefore dismisses concerns that cannot be connected to it. If a proposed moral idea has no connection to properties constitutive of human nature, it has no moral weight.

There is some intuitive plausibility to this peremptory claim. When inequality in resources promotes perfection rather than amusement, it can seem to be morally justified. From a perfectionist standpoint, pleasure and pain can appear to be mere biological signals of good and poor functioning—indicators of what has moral importance, but not significant in themselves. But can one affirm this claim in every case? Can the intrinsic appeal of narrow perfectionism outweigh entirely its omission of other initially attractive ideas?

I am not sure how to resolve this conflict. More specifically, I am not sure whether the most plausible moral theory is a pure perfectionism based on the narrow ideal of human nature or a pluralist view that also gives weight to other moral concerns. There seem to be powerful arguments on either side. Nevertheless, the following seems to me to be true: In any acceptable pluralist theory, perfectionist ideas must play a large role; and among pure moral theories, Aristotelian perfectionism is the most plausible on offer.

Notes

Chapter 1

1. Hamilton, *Lectures on Metaphysics*, p. 14.
2. Rawls, *A Theory of Justice*, p. 325.
3. Green, *Prolegomena to Ethics*, sec. 352.
4. Sidgwick, *The Methods of Ethics*, p. 9.
5. Bradley, "The Limits of Individual and National Self-Sacrifice," pp. 168, 173, 175.
6. Rawls, "The Independence of Moral Theory," p. 5.
7. This division between subjective and objective theories of the good is a division within normative ethics, and independent of the meta-ethical question of whether claims about the good, or moral claims generally, can be "objectively true." This book does not address the meta-ethical question and assumes no view about it.

Chapter 2

1. Plato, *Republic*, 353a.
2. Aristotle, *Nicomachean Ethics*, 1097b33–1098a2.
3. Kant, *The Doctrine of Virtue*, p. 51.
4. Marx and Engels, *The German Ideology*, p. 160.
5. Marx, *Economic and Philosophical Manuscripts*, p. 82; and *Capital*, vol. 1, pp. 283–84.
6. Williams, *Morality: An Introduction to Ethics*, p. 64; see also Nielsen, "Alienation and Self-Realization," pp. 23–24. Williams intends some of his properties, especially the last ones, to be not just morally trivial but repugnant. The issue is tricky, however. If killing things for fun is repugnant, it is primarily because of its effect on the things killed. That the killing is intrinsically evil, or makes the killer's life worse, is a more contentious claim that, in my view, perfectionism need not affirm (see further 2.3.2, 5.2.3, 10.2.3, 10.4.1). In any case, it is sufficient for Williams's objection if his properties are valueless, that is, lack positive worth.
7. Nozick, *Philosophical Explanations*, pp. 515–17.
8. Kripke, *Naming and Necessity*.
9. Hegel, *The Phenomenology of Spirit*, p. 297; and *The Philosophy of Right*, sec. 153.
10. Marx, *Economic and Philosophical Manuscripts*, pp. 83, 89.
11. Nietzsche, *Beyond Good and Evil*, sec. 186; and *The Will to Power*, sec. 693.
12. Other claims that seem equivalent to ones about essence are: that perfection consists in the conformity of human existence with its "idea" or "concept" (Hegel, Marx, Bradley), that something is the "species-being" or "species-activity" of humans (Marx), that some-

thing constitutes "life" or humans' "life-activity" (Marx, Nietzsche), and that certain capacities belong to a human's "real" or "true self," as opposed to his "apparent self" (Kant, Bradley).

13. Another possibility is to apply the distinctiveness test to conjunctions of essential properties. Essential property *A* may not be distinctive of humans, nor essential property *B*, but no other species may have *A* and *B* together. If so, then a further compound view says the conjunction *A & B* belongs to human nature. As initially stated, this view collapses into the simple essence view. If the conjunction *A & B* is unique to humans, it will remain so if essential property *C* is added; species that do not have the smaller conjunction will not have the larger one. So the conjunction of all human essential properties is, as a conjunction, always distinctive. We might try equating human nature with the smallest set of essential properties that is distinctive, but in many cases this collapses into the original essence-and-distinctiveness view. If there is one property that is, individually, both essential and distinctive, the smallest set contains only it.

14. Jean-Jacques Rousseau, *La Nouvelle Héloise,* cinquième partie, lettre 3, quoted in Passmore, *The Perfectibility of Man,* p. 178. See also Nietzsche, *Schopenhauer as Educator,* pp. 1–6; Norton, *Personal Destinies: A Philosophy of Ethical Individualism;* and, for a theory that combines species and individual essences, Unger, *Knowledge and Politics,* pp. 239–40.

15. Aristotle, *Nicomachean Ethics,* 1098a18–20.

16. Aristotle, *Nicomachean Ethics*, 1095b14–1096a10, 1177a11–1179a33; *Eudemian Ethics,* 1215a26–b14; and *Politics,* 1324a24–b1. See also Aquinas, *Summa Theologica,* 2a2ae, q. 182.

17. Could there not be an objective or perfectionist account of well-being, which characterizes well-being not in terms of desires, but in terms of developing human nature? I do not believe there is conceptual room for such an account, for I do not believe "well-being" has any meaning independent *both* of particular accounts of well-being *and* of the moral predicate "good." I do not see that "developing human nature constitutes well-being and is therefore good" says anything over and above "developing human nature is good," and prefer to confine perfectionism to the second, simpler claim.

18. Sidgwick, *The Methods of Ethics,* p. 11.

19. I here assume the modal axiom "If necessarily *p,* then necessarily necessarily *p.*" This axiom characterizes the modal system S4 (and stronger systems) and is widely thought to state an intuitively plausible principle about necessity.

20. Kripke, *Naming and Necessity,* pp. 103–5.

Chapter 3

1. Aquinas, *Summa Theologica,* 1a, q. 5, art. 1; Spinoza, *Ethics,* bk. 2, def. 6.

2. Bradley, *Appearance and Reality,* pp. 360ff; Bosanquet, *The Philosophical Theory of the State,* p. 132.

3. Green, "Liberal Legislation and Freedom of Contract," p. 372; Bradley, *Ethical Studies,* p. 57.

4. Aristotle, *Nicomachean Ethics,* 1097b23–1098a17.

5. Aristotle, *Physics,* 198a26, 199a30-33; *On Generation and Corruption,* 335b6; *Metaphysics,* 1044a36.

6. Marx, *Economic and Philosophical Manuscripts,* p. 80; and *Critique of the Gotha Programme,* p. 569.

7. Aquinas, *Summa Theologica,* 1a2ae, q. 94, art. 3.

8. Nietzsche, *Thus Spoke Zarathustra,* pp. 225–28; *Beyond Good and Evil,* sec. 13; *The Gay Science,* sec. 349; and *The Will to Power,* sec. 650–51.

9. Spinoza, *Ethics,* bk. 3, Definitions of the Emotions, 2–3.

10. Rawls, *A Theory of Justice,* p. 426.

11. Aquinas, *Summa Theologica,* 1a2ae, q. 4, arts. 1, 2; see also q. 2, art. 6, and q. 3, art. 4.

12. Bradley, "Mr. Sidgwick's Hedonism," p. 97.

13. For examples of these interpretations, see Foster, "A Mistake of Plato's in the *Republic*"; Prichard, "The Meaning of *Agathon* in the *Ethics* of Aristotle"; Allen, "The Utilitarianism of Marx and Engels"; and Sidgwick, *Lectures on the Ethics of T. H. Green, Mr. Herbert Spencer, and J. Martineau,* pp. 32–42.

14. Although Rawls's initial formulation of the principle is weak, he later interprets it more strongly. When drawing consequences from it, he says that the tendency it postulates is "relatively strong and not easily counterbalanced," so that in social design "a large place" must be made for it (*A Theory of Justice,* p. 429). For criticism of this stronger principle, see Barry, *The Liberal Theory of Justice,* pp. 27–30; and Haksar, *Equality, Liberty, and Perfectionism,* pp. 194–206.

15. Nietzsche, *Beyond Good and Evil,* sec. 225. As for the view that humans desire pleasure, Nietzsche has nothing but contempt for it and its proponents, especially the "inevitable, indefatigable British utilitarians" (ibid., sec. 228) and "the flathead John Stuart Mill" (*The Will to Power,* sec. 30). "Man does *not* strive for pleasure," he says, "only the Englishman does" (*Twilight of the Idols,* p. 468).

16. Green, *Prolegomena to Ethics,* sec. 361, 276.

17. See Parfit, *Reasons and Persons,* pp. 501–2.

18. For different versions of this view, see Frankena, *Ethics,* pp. 89-92; and my "Virtue as Loving the Good."

19. For recent defences of perfectionist naturalism, see Hampshire, *Thought and Action,* p. 232; Vendler, "The Grammar of Goodness"; Taylor, "Marxism and Empiricism"; Clark, "The Use of Man's Function in Aristotle"; Macpherson, *Democratic Theory: Essays in Retrieval,* pp. 37–38; and Malinovitch, "Knowledge and Evaluation."

20. Given a strong desire doctrine, perfectionism could not do what morality normally does, namely, tell us what goal to aim at. But in telling us how to achieve our one existing goal, it would fill the only action-guiding role that could be filled.

21. Aquinas, *Summa Theologica,* 1a, q. 5, art. 1.

22. Bradley, *Ethical Studies,* p. 65.

23. Ibid., p. 67.

24. Ibid., p. 68.

25. Aristotle, *Physics,* 194a27–30, 195a24–25.

26. MacIntyre, *After Virtue*, chaps. 5 and 7; and Williams, *Ethics and the Limits of Philosophy,* chap. 3. MacIntyre believes a theory similar in some respects to perfectionism is defensible today; Williams does not. Yet, they understand perfectionism in similar ways.

27. Rawls, *A Theory of Justice,* p. 20.

28. Kripke, *Naming and Necessity,* pp. 123–25.

29. Putnam, "Is Semantics Possible?" and "The Meaning of 'Meaning.'"

30. Hempel, *Aspects of Scientific Explanation,* pp. 381–412.

31. See, for example, Brody, *Identity and Essence,* pp. 144–51.

32. Nietzsche, *Beyond Good and Evil,* sec. 36. Note that "intelligible character" is an Idealist synonym for "essence."

Chapter 4

1. Some may deny that lower animals have intentions, as opposed to mere desires. If so, distinctively human psychology begins earlier than I claim.

2. Davidson, *Essays on Actions and Events.*

3. Davidson, "Mental Events," p. 222.

4. This argument would be strengthened if there were evolutionary advantages to rationality, understood as the capacity for sophisticated mental states, and thus evolutionary reasons to expect it in humans. On this topic, see Elster, *Ulysses and the Sirens*, chap. 1.

5. Davidson, "Psychology as Philosophy," pp. 229–38.

6. For a related account of "human" and "person," see Wiggins, *Sameness and Substance*, chap. 6.

7. Another possibility is to define human essential properties statistically, so that "if *most* members of a population lacked them, it would not consist of humans" (Attfield, *A Theory of Value and Obligation*, p. 43 [emphasis mine]). This approach also allows foetuses and the demented to be humans, but it has other unacceptable consequences. First, the statistical approach leaves it unclear what determines membership in the human species. We cannot say that a being is a human if and only if it has the properties essential to humans: Some humans may lack these properties. So what makes a human a human? Second, the approach cannot treat being a human as essential to individual humans. Imagine that humans are essentially rational, in the statistical sense, and that *A* is a rational human. There is a possible world in which a *B* exists who shares all of *A*'s intrinsic properties and also shares *A*'s history, but in which most other members of *B*'s species are not rational. On the statistical view, the beings in *B*'s world are not humans, and therefore *B* is not a human. But *B* is surely identical to *A:* the two are a perfect match. And this implies that *A* is not essentially a human. Being a human is just another of *A*'s properties, like being a lawyer or a hockey fan, with no special importance for *A*'s identity (2.1.2).

8. See Becker, "Human Being: The Boundaries of the Concept," p. 355.

9. It may be objected that a brain-dead body is not a human because its heart and lungs, although functioning, are not doing so autonomously, but this objection has the absurd consequence that people on kidney machines or wearing pacemakers are not humans. See Green and Wikler, "Brain Death and Personal Identity," pp. 108–9.

10. Given the nature of sophisticated rationality, there may be no one precise moment when it appears, but instead a gradual process. If so, there will be three categories of beings: those that are clearly almost-humans (early foetuses), those that are clearly humans (adults), and those that have an indeterminate status between the two (newborn babies?).

11. It may be objected that, if babies are not humans, our duties concerning them are duties to bring a future human into existence and thus no different in kind from the presumably weak duty to conceive humans. But there can be differences in degree. A newborn baby's potential for human life is much closer to realization than is that of unfertilized cells and may therefore generate stronger duties.

12. My claim is not that species can have no relational essential properties (see, to the contrary, 8.4.2), just that the specific relational properties central to evolutionary explanations are not plausibly among them.

13. A similar argument tells against the proposal, based on some biological practice, to define species boundaries and essences by the capacity for interbreeding. On this proposal a sterile human, one that cannot breed, is not a human, which is absurd. Facts about reproduction, although crucial for determining individual essences, are irrelevant to questions about the generic human essence. This is why I do not include the reproductive system as one of those that define our physical essence (4.1.1).

Chapter 5

1. Aristotle, *Nicomachean Ethics,* 1177b35, 1169a6–8; *Politics,* 1323b7–34.

2. Humboldt, *The Limits of State Action,* p. 16.

3. Bradley, *Ethical Studies,* pp. 228, 215.

4. See Slote, *Common-Sense Morality and Consequentialism,* chap. 3.

5. A different satisficing view gives each human a duty to pursue perfection not up to some absolute threshold that is the same for all, but to some reasonable fraction of the greatest perfection possible for him. This view does give Mozart a duty to develop his musical talents, although not a maximizing duty: he must seek, for example, three quarters of the greatest musical achievement possible for him. But this view is still not as plausible, given perfectionist values, as maximizing.

6. See Finnis, *Natural Law and Natural Rights,* chap. 5. Finnis argues against maximizing that it is impossible to weigh against each other the different elements of an ideal (pp. 112–15). For one account of this weighing, see chapter 7 in this book.

7. If a theory uses a concept of right action tied to actual consequences, it can make distinctions in cases involving risk. But it still cannot distinguish *A*'s and *B*'s acts in the first example, where their options differ.

8. Aristotle, *Nicomachean Ethics,* 1100b20–22, 1100b30–1101a6; *Politics,* 1332a19–21.

9. Aristotle, *Nicomachean Ethics,* 1099a31–b3.

10. Aristotle, *Politics,* 1283a35–37, 1331b39–41, 1332a39–b2.

11. Aristotle, *Nicomachean Ethics,* 1100b22–1101a8.

12. Aristotle, *Nicomachean Ethics,* 1095b25–26, 1178a24–b7; *Politics,* 1323b22–30. At *Nicomachean Ethics,* 1097b7–22, Aristotle calls perfection "self-sufficient" in another sense, meaning that someone who has it lacks nothing that is good. This is not the sense I intend here.

13. Kant, *Religion Within the Limits of Reason Alone,* p. 36.

14. Greenwood, *Aristotle's Nicomachean Ethics Book Six,* pp. 46–47.

15. The distinctions in these paragraphs come from Parfit, *Reasons and Persons,* p. 143.

16. Aristotle, *Nicomachean Ethics,* 1098a18–20, 1095b14–1096a10, 1177a11–1179a33; *Eudemian Ethics,* 1215a26–b24; and *Politics,* 1324a24–b1. See also Aquinas, *Summa Theologica,* 2a2ae, q. 182.

17. Parfit, *Reasons and Persons,* pp. 165–66.

18. See ibid., part II.

19. See my "Consequentialism and Content," pp. 73–74.

20. Nagel argues that if someone desires to climb Mt. Kilimanjaro or to play all the Beethoven piano sonatas, this desire makes his doing so good "for him," but not agent-neutrally (*The View From Nowhere,* pp. 166–71). But Nagel's argument concerns only the value (allegedly) conferred by desires and assumes "that we can abstract from any intrinsic value the achievement may have which does not depend on [the person's] interest at all" (ibid., p. 169). Unlike Nagel's, my argument does concern the intrinsic value of the achievement, which it claims, consistently with Nagel, is best understood as agent-neutral.

21. Kant, *The Doctrine of Virtue,* pp. 44–45.

22. Kant seems to think it is: see *The Critique of Practical Reason,* pp. 155–65; and *The Doctrine of Virtue,* pp. 146–60.

23. Aristotle, *Nicomachean Ethics,* 1094b7–10.

24. Aristotle, *Politics,* 1279a17–22, 1283b40–44, 1277b16–1279b8; also *Nicomachean Ethics,* 1129b25–1130a7. I do not call Aristotle's perfectionism agent-neutral because it shows no concern for those outside one's city-state.

25. Aristotle, *Nicomachean Ethics,* 1130a4. The phrase comes from Plato, *Republic,* 343c.

26. Aristotle, *Nicomachean Ethics,* 1129b15–19.

27. Green, *Prolegomena to Ethics,* sec. 232, 281; see also 240.

28. Sidgwick, *Lectures on the Ethics of T. H. Green, Mr. Herbert Spencer, and J. Martineau,* p. 67.

29. Despite its implausibility, absolute non-competition continues to be affirmed. See Adler, *The Time of Our Lives,* pp. 172–74; Norton, *Personal Destinies: A Philosophy of Ethical Individualism,* p. 14; and Unger, *Knowledge and Politics,* pp. 247, 260, 269. Claims that, although not absolute, are still very strong are in Nozick, *Philosophical Explanations,* pp. 510–15; and Macpherson, *Democratic Theory: Essays in Retrieval,* pp. 54–55.

30. Aristotle, *Politics,* 1263a25–29.

31. Nietzsche, *Ecce Homo,* in *On the Genealogy of Morals/Ecce Homo,* p. 234.

32. Nietzsche, *The Gay Science,* sec. 338. See also *Beyond Good and Evil,* sec. 44, 61.

33. Green, *Prolegomena to Ethics,* sec. 245.

34. Marx and Engels, *The Communist Manifesto,* in *Karl Marx: Selected Writings,* p. 326.

Chapter 6

1. For the contrary view, see Griffin, *Well-Being: Its Meaning, Measurement and Moral Importance,* pp. 104–5.

2. If a theory sums across both times and persons, the order in which it does so is unimportant; but if it uses almost any other aggregative principles, order is vital. For an illustration, see my "Average Utilitarianisms."

3. McTaggart, *The Nature of Existence,* vol. 1, pp. 452–53. For the version across persons, see Parfit, *Reasons and Persons*, pp. 387–90; and "Overpopulation and the Quality of Life," pp. 148–51.

4. Sikora, "Is It Wrong to Prevent the Existence of Future Generations?" p. 116; and Parfit, *Reasons and Persons,* p. 420.

5. For literary expressions of this view, see Swift's account of the *struldbrugs* in *Gulliver's Travels,* part III, chap. 10; and Housman, "To an Athlete Dying Young," in his *Complete Poems*, pp. 32–33.

6. Xenophon, *Memorabilia,* bk. 4, chap. 8, sec. 1, 5–8; and *Apology,* sec. 5–9.

7. Plato, *Republic,* 387b, 486ab; see also the discussion of medical practices at 406–10.

8. Nietzsche, *Thus Spoke Zarathustra,* pp. 183–85.

9. The first suggestion recurs in similar language elsewhere in Nietzsche. See *The Gay Science,* sec. 281; *Twilight of the Idols,* pp. 536–37; and *The Will to Power,* sec. 864 (but contrast *The Use and Abuse of History,* pp. 53–54). The second is harder to find unambiguously stated, just as it is less clearly present in the quoted passage.

10. Aristotle, *Nicomachean Ethics,* 1115a26–27; see also 1166a19, 1168a5–6, and 1170a26–28.

11. Ibid., 1117b10–12; but contrast 1169a22–25.

12. Ibid., 1098a17–19; see also *Eudemian Ethics,* 1218b35–39.

13. It may be thought that averaging perfectionism also requires us to kill those who are less perfect than the average in their population. This is not so if the theory aggregates across times before persons: It then tells us to maximize the average lifetime value per life in history, and once a life has begun it cannot be removed from the calculations relevant to this average.

An averaging theory that calculates first within lives tells us not to create people whose lives will be below average, but it cannot tell us to kill them once their lives have begun.

14. Nietzsche, *Schopenhauer as Educator,* pp. 59–60.

15. Nietzsche, *The Will to Power,* sec. 766. See also *The Use and Abuse of History,* p. 59; *The Gay Science,* sec. 23; *Beyond Good and Evil,* sec. 126, 199, 258; *On The Genealogy of Morals,* pp. 56, 124–25; and *The Will to Power,* sec. 660, 681, 877, 987, 997.

16. Nietzsche, *The Will to Power,* sec. 373; see also *Beyond Good and Evil,* sec. 265.

17. Nietzsche, *Mixed Opinions and Maxims,* pp. 174–75.

18. Compare G. H. Hardy on the greatness of mathematicians: "The real tragedy about Ramanujan was not his early death. It is of course a disaster that any great man should die young, but a mathematician is often comparatively old at thirty, and his death may be less of a catastrophe than it seems. Abel died at twenty-six and, although he would have added a great deal more to mathematics he could hardly have become a greater man" (*Ramanujan: Twelve Lectures on Subjects Suggested by His Life and Work,* p. 6).

19. Kaufmann, *Nietzsche: Philosopher, Psychologist, Antichrist,* p. 150. The textual basis for Kaufmann's reading seems to be *The Will to Power,* sec. 53.

20. Nietzsche, *The Will to Power,* sec. 693.

21. Nietzsche, *Beyond Good and Evil,* sec. 258.

22. Compare Rawls's "lexical difference principle," *A Theory of Justice,* pp. 82–83. Lexical maximax differs from maximax in cases where we cannot affect the best humans but can improve the lives of some less good humans. Maximax says there is no reason to do this; lexical maximax says there is.

23. We could combine maximax with a perfectionist egalitarian principle, but the result would be strange: two opposed principles governing the same substantive values. The combination of maximax with a distributively neutral principle is better represented by the finite priority principle.

24. See Gordon and White, *Philosophers as Educational Reformers: The Influence of Idealism on British Educational Thought and Practice.*

25. Nagel, "The Fragmentation of Value," p. 132.

26. Ibid., p. 130; see also Nagel, *Equality and Partiality,* chap. 12.

27. Parfit, "Overpopulation and the Quality of Life," pp. 161–64.

28. See Parfit, *Reasons and Persons,* part IV.

Chapter 7

1. Aristotle, *Nicomachean Ethics,* 1177b26–1178a7; see also *Eudemian Ethics,* 1249b6–25.

2. According to Aquinas, the perfection of our bodily parts is necessary to the blessed state that will follow the resurrection of the body (*Summa Theologica,* 1a2ae, q. 3, art. 3; q. 4, art. 5; see also Supp., qq. 82–84), and also characterized the slightly less perfect state that preceded the Fall. Arguing against Origen that humans would indeed have reproduced by sexual intercourse had they stayed longer in a state of innocence, Aquinas adds that this intercourse would have been vastly more pleasurable than anything we experience today because of "the greater purity of nature and the greater sensibility of the body" (ibid., 1a, q. 98, art. 2).

3. Ibid., 2a2ae, q. 152, arts. 2, 3.

4. A theory that takes this line need not require us always to pursue the highest excellence in ourselves. If it is agent-neutral (5.2.3), it can tell us to sacrifice some of our own achievement of this good to promote it in others. This seems to be Aquinas's view. Despite agreeing

with Aristotle that contemplation is in itself better than any practical good (*Summa Theologica,* 2a2ae, q. 182, art. 1), Aquinas finds it meritorious to abandon one's own contemplation in order to promote it in others (ibid., 2a2ae, q. 182, art. 2, q. 184, art. 7; see also 3a, q. 40, art. 1, ad 1). Because "it is better to illumine than merely to shine" (ibid., 2a2ae, q. 188, art. 6), the most praiseworthy lives are devoted, as Aquinas's was, to teaching as well as study.

5. Aristotle, *Nicomachean Ethics,* 1177b26–1178a8.

6. Ibid., 1177a21.

7. For these three arguments see Aristotle, *Nicomachean Ethics,* 1177a24–b25; and Aquinas, *Summa Theologica,* 2a2ae, q. 182, art. 1.

8. See van Fraassen, "Presuppositions, Supervaluations, and Free Logic"; and Sen, *Collective Choice and Social Welfare,* chap. 7.

9. Some may say that, even using supervaluations, we cannot go far measuring perfectionist values. The acceptable principles for comparison, aggregation, and so on are so diverse that even together they leave vast areas where no determinate claims can be made. This argument is not technical but moral. It concerns how many principles our intuitive judgements rule out and is an argument I reject on moral grounds. Although we cannot approach precise cardinality, there are often general conditions on measurement that we can formulate and choose among. This claim cannot, however, be defended in the abstract. Its plausibility will emerge, I hope, in the detailed discussions of chapters 6 through 10.

10. For philosophical uses of indifference graphs, see Barry, *Political Argument,* pp. 3–8; and Rawls, *A Theory of Justice,* pp. 37–40.

11. The finite priority principle (6.2.1) holds that the absolute value of a gain in perfection increases the more perfection one has, so twenty units each of two perfections are more than twice as good as ten each. I assume that we have rejected this view, partly because of its anti-egalitarian consequences.

12. Hamilton, *Lectures on Metaphysics,* p. 14.

13. Humboldt, *The Limits of State Action,* p. 16.

14. Rashdall, *The Theory of Good and Evil,* vol. 1, pp. 220–21. Elsewhere Rashdall says that "no amount of one good can compensate for the absence or deficiency of the other" (vol. 2, p. 39), an even clearer balancing claim.

15. Humboldt, *The Limits of State Action,* p. 16.

16. Marx and Engels, *The German Ideology,* p. 169; see also Marx, *Grundrisse,* pp. 325, 487–88.

17. Nietzsche, *Thus Spoke Zarathustra,* p. 250; *Beyond Good and Evil,* sec. 204, 205, 212, 257; *On the Genealogy of Morals,* p. 147; *Twilight of the Idols,* p. 508; *The Antichrist,* p. 647; and *The Will to Power,* sec. 390, 881.

18. Nietzsche, *Beyond Good and Evil,* sec. 212.

19. It is this possibility—different returns of perfection at different levels, as well as to people with different talents—that forces us to equate one unit of perfection with what an *average* person can achieve in a given period of time (7.2.1). With precise cardinal measures of the individual excellences we can compare goods in one area, for example, Leonardo's paintings with a beginner's. But we need to normalize these measures, and do so by considering an average person's achievements in a fixed time.

20. Rashdall, *The Theory of Good and Evil,* vol. 2, p. 64. How this argument is supposed to mesh with Rashdall's own remarks about "harmonizing" different elements (see n. 14), I do not venture to say.

21. Darwin, "Memories of Down House," pp. 119–20.

22. Plato, *Republic,* 503c.

23. Humboldt, *The Limits of State Action,* p. 17.

Chapter 8

1. Bentham, *An Introduction to the Principles of Morals and Legislation,* chap. 4.

2. Aquinas, *Summa Theologica,* 1a, q. 48, arts. 1–2; Leibniz, *Theodicy,* pp. 135–36, 140–41, 352.

3. Aristotle is not entirely explicit about this, but his talk of theory always concerns "contemplating" and "activity"; see, e.g., *Nicomachean Ethics,* 1177a12–18, b3. For Aquinas see *Summa Theologica,* 1a2ae, q. 5, art. 4; 2a2ae, q. 182, art. 3.

4. Aquinas, *Summa Theologica,* 1a2ae, q. 3, art. 2.

5. The dispositional view may seem to have its own counter-intuitive consequence: that someone can have high theoretical value while asleep or in a coma. We could avoid this consequence by distinguishing between grades of dispositional belief. When a person is asleep, a dispositional belief she has is two steps from being occurrent: the person must first wake up, then attend to that belief. When she is awake, her belief is just one step from being occurrent. This may be a reason to give the first kind of dispositional belief less weight, but I am not sure that this emendation is necessary. Would we lower our estimate of the value of, say, Newton's knowledge if we learned that he slept an hour a day more than we thought?

6. Both aphorisms are quoted in Hamilton, *Lectures on Metaphysics,* p. 9.

7. More subtly, we can make the value of an additional state in a kind get smaller the more states a person has. Thus, the first belief in a kind may count as one perfectionist state, the second as 0.5 of a state, the third as 0.25, and so on. (In each case, this fraction is multiplied by a standard measure of quality.) This places a finite upper bound on measures of number while giving people some credit for believing the logical consequences of their beliefs.

8. Browning, "Bishop Blougram's Apology," lines 491–93, and "Andrea del Sarto," lines 97–98, both in *Poems of Robert Browning.*

9. Aristotle, *Nicomachean Ethics,* 1144a7–11; see also Aquinas, *Summa Theologica,* 1a2ae, q. 57, art. 5.

10. Note that the externalist model has none of these consequences. Often one of a pair of inconsistent beliefs derives from a reliable method, and sometimes both do.

11. See Gibson, "Rationality," pp. 214–15.

12. Aristotle, *Nicomachean Ethics,* 1146a6–7, 1152a5–8; Nietzsche, *The Will to Power,* sec. 46.

13. This trichotomy cuts across the traditional division between internal and external goods (Plato, *Euthydemus* 279ab, *Philebus* 48e, *Laws* 343e; Aristotle, *Nicomachean Ethics,* 1098b12–13). Knowledge is usually counted as an internal good and wealth as external, but both true beliefs and material possession involve a relation between the mind and something outside it. For a clear illustration of the trichotomy, see Moore's discussion of aesthetic values in *Principia Ethica,* pp. 188–98.

14. Nozick, *Anarchy, State, and Utopia,* pp. 42–45; and Glover, *What Sort of People Should There Be?* chaps. 7–8.

15. Neither attempt view leaves perfection entirely in a person's control. On both views, the quality of the states he can form depends on his native abilities, which he cannot change. And on deserving attempt, what he is justified in believing depends on his evidence. So neither attempt view makes perfection entirely self-sufficient or removes entirely the reason for making perfectionism consequentialist (5.1.3).

16. In chapter 2 I rejected views like the distinctiveness view, which make our nature depend on facts about other species. This does not undermine the argument here: That the concept of nature must be non-relational does not imply that properties falling under it must be non-relational. And in chapter 4 I argued, against some claims based on evolutionary biology,

that the properties highlighted by this biology are often relational. The concern there, however, was not with relationality in general, but with the specific relational properties central to evolutionary explanations. It is implausible that a butterfly species is essentially yellow against a yellow background (4.4.1), but not implausible that humans necessarily have some true beliefs about their surroundings.

17. This idea is beautifully expressed by Lockwood: "Set against an ideal of human life as a meaningful whole, we can see that premature death can, as it were, make nonsense of much of what has gone before. Earlier actions, preparations, planning, whose entire purpose lay in their being directed towards some future goal, become, in the face of an untimely death, retrospectively pointless—bridges, so to speak, that terminate in mid-air, roads that lead nowhere" (Lockwood, "Singer on Killing and the Preference for Life," p. 167).

18. Kant, *Foundations of the Metaphysics of Morals,* p. 10.

19. Aristotle, *Nicomachean Ethics,* 1144a26; see also 1142b16–35 and the analogy between excellence and winning at the Olympic Games at 1099a3–5.

20. Ibid., 1100a10–31, 1101a21–b9.

21. Sidgwick, *The Methods of Ethics,* pp. 398–407; Bradley, "Mr. Sidgwick's Hedonism," p. 95; Green, *Prolegomena to Ethics,* sec. 364; McTaggart, *Studies in Hegelian Cosmology,* p. 79; and Rashdall, *The Theory of Good and Evil,* vol. 1. pp. 68–69.

Chapter 9

1. Humeans may balk at this inference. They may agree that someone with more sophisticated mental states exercises rationality more, but deny that the person is therefore more rational. Not much turns on this terminological issue. The important Aristotelian claim is that humans *ought* to exercise rationality more, which no Humean theory says.

2. To prevent some items of knowledge, for example, scientific laws, from having infinite extent, we should make the value of additions to objects or times get less the more of them there are, and diminish towards zero. (This move is plausible in itself.) We also need to be able to calculate the extent of logically complex contents, such as *p or q, not-r,* and *possibly-s*. I have no complete account of these, but in truth-functional cases their extent can perhaps equal that of the least extended state that makes them true: thus, the less extended of *p* and *q* for *p or q,* the least extended falsifier of *r* for *not-r,* and so on.

3. Although number conditions limit the way value moves up a hierarchy, value at the bottom of a structure is safe from defects above. Someone may have beliefs *s* and *t* that are explained by *q* and also believe that *q* is explained by *p* (see Figure 9.1). If *p* is false, this negates *p*'s quality, which is considerable. But it does not affect *q,* which retains the hierarchy below it, nor does it affect *s* and *t*. Similarly, someone may achieve several subordinate ends as means to a final end but find that they do not bring it about. This robs his action of much value, namely, that in his final end, but it leaves his subordinate ends and any lesser hierarchies they form.

4. Imagine that one person uses *p* to explain *q* and *r,* which are both about hockey, whereas another uses *p* to explain *q* and *s,* where *s* is about golf. The first person's explanatory hierarchy contains *p*-truths, *r*-truths, and hockey-truths, a total of three different kinds. The second's contains *p*-truths, *s*-truths, hockey-truths, and golf-truths, a total of four kinds. In the everyday sense, the second person uses *p* to explain more kinds of truth.

5. This is not the only way to calculate dominance. A state's dominance could equal the sum of the dominance values of the states subordinate to it, the sum of the squares of those values, or some other mathematical function. As elsewhere, our informal judgements of dominance should be seen as arising from supervaluations over many different precise measures (7.1.3).

6. Harman, "Practical Reasoning," p. 462.

7. These styles are associated with, respectively, Athens and Manchester, in Dyson, *Infinite in All Directions,* chap. 3.

8. Is this argument consistent with having number and dominance count kinds (8.1.4, 9.2.2)? Are not the extra truths that make for precision all of the same kind? I think not. What we must count together, to avoid infinite measures, are trivial logical consequences and replications of the same property at different places and times. With precise knowledge, however, we have slightly different properties at the same place and time. (There must be just a finite number of such properties, but the division into kinds can ensure this.) In any case, the best measures of number and dominance give additional states in a kind some (although diminishing) value (see chap. 8, n. 7), which also allows some value to precision.

9. When it is calculated using kinds that can include each other, dominance measures what is sometimes called "organic unity"; see Nozick, *Philosophical Investigations,* pp. 415–28, 724.

10. Sidgwick, *The Methods of Ethics,* p. 381.

11. Rawls, *A Theory of Justice,* p. 426.

12. Herbert Read, quoted in Gardner, *Frames of Mind: The Theory of Multiple Intelligences,* p. 188.

13. "Shown a photograph of a nondescript instant on the ice, Gretzky can replace the unpictured performers here and there about the periphery and usually recall what became of them the next second. Glancing at the basketball photo in the morning paper, [Larry] Bird's automatic thought, essentially a reflex, is to note approximately what time the photographer had to snap his picture to make the deadline" (Callahan, "Masters of Their Own Game," p. 38).

14. Sturt, *The Wheelwright's Shop,* pp. 19–20, 24.

15. Herrigel, *Zen in the Art of Archery.* Nietzsche, too, thinks consciousness during an activity is a mark of imperfection; see *Beyond Good and Evil,* sec. 191; *On the Genealogy of Morals,* pp. 39, 84–87; *Ecce Homo,* pp. 253–55; and *The Will to Power,* sec. 289.

16. For an insightful and delightful account of games, see Suits, *The Grasshopper: Games, Life, and Utopia.*

Chapter 10

1. See Mill, *Considerations on Representative Government,* pp. 193, 195, 203–5.

2. See Barber, *Strong Democracy: Participatory Politics for a New Age,* chap. 10.

3. This uses within the measure of extent something parallel to the "marginal consequences" interpretation of consequentialism; see Regan, *Utilitarianism and Co-operation,* pp. 13–17; and Parfit, *Reasons and Persons,* pp. 67–70.

4. Bradley, *Ethical Studies,* essay 5.

5. Nozick, *Philosophical Explanations,* pp. 464–65.

6. See my "Virtue as Loving the Good."

7. Nietzsche, *The Will to Power,* sec. 877.

8. Aristotle, *Nicomachean Ethics,* 1098b34–36.

9. Williams, *Morality: An Introduction to Ethics,* pp. 65–66.

10. Rawls, *A Theory of Justice,* p. 450.

11. It may be objected that number is reduced if we fail to have detailed intentions about the future. This is true: Without such intentions, we have fewer value-bearing states in the present. But if forming these intentions would prevent our pursuing higher-quality goals in the present, as it surely would, our loss of perfection is usually made good.

12. See, e.g., Williams, "Persons, Character and Morality."

13. Aristotle, *Nicomachean Ethics,* 1147b3–5.

14. Aristotle, *Eudemian Ethics,* 1214b6–10.

15. Aristotle, *Nicomachean Ethics,* 1111b7–15.

16. Ibid., 1105a10.

17. Ibid., 1129b25-1130a13; *Politics*, 1277b25–29.

18. Aristotle, *Metaphysics,* 1025b1–1026a33. See also Aquinas, *Summa Theologica,* la–2ae, q. 3, art. 8.

19. Leibniz, *Principles of Nature and of Grace,* sec. 13.

20. Leibniz, *The Leibniz-Arnauld Correspondence,* p. 15.

21. Nietzsche, *The Gay Science,* sec. 290.

22. Nietzsche, *The Will to Power,* sec. 46.

23. Nietzsche, *Beyond Good and Evil,* sec. 188. See also *Thus Spoke Zarathustra,* p. 188; *Beyond Good and Evil,* sec. 19, 208; *On the Genealogy of Morals,* p. 16; and *The Will to Power,* sec. 334, 387. For a similar reading of Nietzsche, see Nehamas, *Nietzsche: Life as Literature,* chap. 6.

24. Nietzsche, *The Will to Power,* sec. 962; and *Beyond Good and Evil,* sec. 208. See also *The Gay Science,* sec. 356; *Thus Spoke Zarathustra,* pp. 156, 320; *Beyond Good and Evil,* sec. 257; *On the Genealogy of Morals,* pp. 58-60; *Twilight of the Idols,* p. 543; *The Will to Power,* sec. 65, 527.

25. Nietzsche, *On the Genealogy of Morals*, p. 87; see also p. 86.

26. Bradley, *Appearance and Reality,* p. 370.

27. Ibid., p. 148.

28. Ibid., p. 363. See also Bradley, *Essays on Truth and Reality,* pp. 114, 202, 211, 227n., 241; *The Principles of Logic,* pp. 684–88; and Ross, *The Right and the Good,* pp. 139, 147–49.

29. Bradley, *Appearance and Reality,* pp. 414–15.

Chapter 11

1. Aristotle, *Politics,* 1252b29–30, 1324a23–25; see also 1277b16–1279b8, 1283b40–44, 1323a17–19, 1325a7–10.

2. Nietzsche, *The Will to Power,* sec. 660; see also *Beyond Good and Evil,* sec. 257.

3. Marx, *Capital,* vol. 1, p. 284.

4. Humboldt, *The Limits of State Action,* p. 19.

5. Ibid., p. 16

6. Mill, *On Liberty,* p. 116.

7. Ibid., p. 115.

8. Griffin, *Well-Being: Its Meaning, Measurement and Moral Importance,* pp. 243–45.

9. Although autonomy realizes agency, it is not identical to agency because it involves only intentions about one's own life. Determining what is and is not true about other people or the material world involves a form of agency distinct from autonomy—call it *external power*.

10. Green, *Prolegomena to Ethics,* sec. 332.

11. Green, "Liberal Legislation and Freedom of Contract," p. 374.

12. Green, *Lectures on the Principles of Political Obligation,* sec. 209; see also sec. 18.

13. Nettleship, *Lectures on the Republic of Plato,* p. 144.

14. Barker, *The Political Thought of Plato and Aristotle,* p. 246. See also Bosanquet, *The Philosophical Theory of the State,* pp. xxv-xxvi, 190–93; and Hobhouse, *Liberalism,* pp. 111–13, 121–23, 143.

15. Green, *Lectures on the Principles of Political Obligation,* sec. 16; see also sec. 209.

16. Humboldt, *The Limits of State Action,* pp. 28–29.

17. Rashdall, *The Theory of Good and Evil,* vol. 1, pp. 297–99.

18. Humboldt, *The Limits of State Action,* p. 25.

19. Rawls, *A Theory of Justice*; Dworkin, "Liberalism"; and Ackerman, *Social Justice in the Liberal State.*

20. Mill, *On Liberty,* p. 73; also p. 133. Humboldt does endorse neutrality (*The Limits of State Action,* p. 25), but only by stretching his arguments beyond plausibility.

Chapter 12

1. Richards, "Justice and Equality," pp. 245–46; and Haksar, *Equality, Liberty, and Perfectionism,* p. 54.

2. Does the objection apply at least to maximax perfectionism (6.2)? It is not clear. Many partisans of deep equality cite Rawls's theory of justice, with its maximin distributive principle, as a paradigm expression of equal concern. But if there is a sense in which maximin gives the well-off *equal* consideration with the badly-off, there must be a sense in which maximax also gives equal consideration.

3. Rawls, *A Theory of Justice,* pp. 137–42; Ackerman, *Social Justice in the Liberal State;* Dworkin, "Liberalism," pp. 127–28.

4. Rawls, "Kantian Constructivism in Moral Theory," p. 542; and "Justice as Fairness: Political not Metaphysical," p. 230. See also Nagel, "Moral Conflict and Political Legitimacy"; and Larmore, *Patterns of Moral Complexity,* chap. 3.

5. Ackerman, *Social Justice in the Liberal State,* pp. 368–69.

6. Rawls, "Justice as Fairness: Political not Metaphysical," p. 230.

7. Rawls, "Kantian Constructivism in Moral Theory," pp. 525, 534.

8. Rashdall, *The Theory of Good and Evil,* vol. 1, pp. 238–39. On a charitable reading Rashdall is advocating, not death for Asians and Africans, just limits on their procreation.

9. Aristotle, *Nicomachean Ethics,* 1131a24–29.

10. Rashdall, *The Theory of Good and Evil,* vol. 1, p. 263; see also pp. 251–53.

11. Green, *Prolegomena to Ethics,* sec. 381; see also sec. 270.

12. Rashdall, *The Theory of Good and Evil,* vol. 1, p. 250; see also pp. 229, 234.

13. Aristotle, *Politics,* 1254b2–22, 1259b21–1260b8.

14. Ibid., 1278a20–21, 1328b38–1329a1.

15. Block and Dworkin, "IQ: Heritability and Inequality, Part 1," pp. 394-407; and Gould, *The Mismeasure of Man*, chap. 6.

16. Gardner, *Frames of Mind: The Theory of Multiple Intelligences,* pp. 63–67 and *passim.*

17. I happen to be good at abstract reasoning and fairly good with words, but I am hopeless spatially. I was a consistent loser at chess; despite playing hockey for years I still cannot judge a simple carom of the puck off the boards; and if given the map on a wilderness canoe trip, I will have my group lost in half an hour. I once accomplished this feat while canoeing down a river.

18. Rawls, *A Theory of Justice,* p. 330.

19. This example comes from Murray, *Losing Ground: American Social Policy 1950–1980,* p. 155. For the same argument, see Harrington, *The New American Poverty,* pp. 202–6.

20. Lewis, *The Children of Sanchez,* esp. pp. xxvi–xxvii; Harrington, *The Other America;* and Banfield, *The Unheavenly City.*

21. Lewis, *The Children of Sanchez,* pp. 425, 477, 499.

22. Plato, *Republic,* 421de; see also 556bc; and *Laws,* 742e–744e.

23. Aristotle, *Politics,* 1256b27–1257a18, 1257b30–1258a17, 1295b2–11; also *Eudemian Ethics,* 1249a21–b5.

24. Rashdall, *The Theory of Good and Evil,* vol. 1, p. 272; also 258n., 259. See also de Jouvenel, *The Ethics of Redistribution,* pp. 22, 40, 63; and Rawls, *A Theory of Justice,* p. 290.

25. Plato, *Republic,* 372c–373e.

26. See Cohen, "Robert Nozick and Wilt Chamberlain: How Patterns Preserve Liberty," p. 258; and Nielsen, *Equality and Liberty: A Defense of Radical Egalitarianism.*

Chapter 13

1. There might be a reason to give more resources to the less talented, but this inequality is benign, not invidious. It reduces rather than increases inequality in what truly matters in life.

2. Marx and Engels, *The Holy Family,* p. 154. The passage reports the views of the French materialists, but Marx and Engels are clearly sympathetic, concluding that materialism is "necessarily . . . connected with communism and socialism."

3. Marx and Engels, *The Communist Manifesto,* p. 140.

4. Does Marx believe in diminishing marginal utility? His talk of "needs" suggests priority for that minimum of resources that makes for security and independence, but he shows no sympathy for the idea that, once total production has reached a certain level, it matters less that it continue to increase beyond that level. Questions about the sources of Marx's egalitarianism are difficult, partly because it is not clear how much derives from his perfectionism and how much from independent claims about distributive justice.

5. For recent work on these "New Liberal" writers see Freeden, *The New Liberalism: An Ideology of Social Reform;* Clarke, *Liberals and Social Democrats;* and Collini, *Liberalism and Sociology: L. T. Hobhouse and Political Argument in England 1880–1914.*

6. Hobhouse, *Liberalism,* pp. 111, 122–23; see also p. 132; *The Rational Good,* pp. 20–21, 39–40, 100; and *The Elements of Social Justice,* pp. 23ff., 67–70.

7. Hobhouse, *Liberalism,* pp. 121–22; see also pp. 111–13, 123, 143.

8. Ibid., pp. 201n., 174–76, 176–80.

9. Ibid., p. 128.

10. Ibid., pp. 163–64.

11. Hobhouse, *The Elements of Social Justice,* p. 109.

12. Hobhouse, *Liberalism*, pp. 128–29.

13. Tawney, *Equality,* p. 43; see also p. 90.

14. Ibid., p. 49.

15. Ibid., p. 81.

16. Ibid., p. 87.

17. Hegel, *The Philosophy of Right,* sec. 41–57.

18. Green, *Lectures on the Principles of Political Obligation,* sec. 214, 213; see also sec. 217, 220, 221.

19. Bosanquet, "The Principle of Private Property," p. 310; see also pp. 308–9, 311; and *The Philosophical Theory of the State,* pp. 302–4.

20. Bosanquet, "Luxury and Refinement," p. 286. See also Scruton, *The Meaning of Conservatism,* pp. 98–102, 125–29.

21. Green, *Lectures on the Principles of Political Obligation,* sec. 223–25. See also Bosanquet, "The Antithesis between Individualism and Socialism Philosophically Considered," pp. 315–16; and *The Philosophical Theory of the State,* p. 193.

22. Bosanquet, *Aspects of the Social Problem,* pp. vii–viii; see also "Character in Its Bearing on Social Causation," p. 111; and "Luxury and Refinement," pp. 295–96.

23. Bosanquet, *The Philosophical Theory of the State,* p. ix; see also pp. 288–92.

24. Hobson, *Problems of Poverty,* p. 12.

25. Cohen, *Karl Marx's Theory of History: A Defence,* chap. 11; see also Schor, *The Overworked American: The Unexpected Decline of Leisure.*

26. Bell, *The Cultural Contradictions of Capitalism.*

Bibliography

Ackerman, Bruce A. *Social Justice in the Liberal State.* New Haven: Yale University Press, 1980.

Adler, Mortimer J. *The Time of Our Lives.* New York: Holt, Rinehart and Winston, 1970.

Allen, D. P. H. "The Utilitarianism of Marx and Engels," *American Philosophical Quarterly* 10 (1973): 189–99.

Aquinas, St. Thomas. *Summa Theologica.* Translated by the Fathers of the English Dominican Province. 4 vols. Westminster, Md.: Christian Classics, 1981.

Aristotle. *Eudemian Ethics.* Translated by J. Solomon, in *The Complete Works of Aristotle* edited by Jonathan Barnes. 2 vols. Princeton: Princeton University Press, 1984.

———. *Metaphysics.* Translated by W. D. Ross. In *The Complete Works of Aristotle,* edited by Jonathan Barnes. 2 vols. Princeton: Princeton University Press, 1984.

———. *Nicomachean Ethics.* Translated by W. D. Ross. In *The Complete Works of Aristotle,* edited by Jonathan Barnes. 2 vols. Princeton: Princeton University Press, 1984.

———. *On Generation and Corruption.* Translated by H. H. Joachim. In *The Complete Works of Aristotle,* edited by Jonathan Barnes. 2 vols. Princeton: Princeton University Press, 1984.

———. *Physics.* Translated by R. P. Hardie and R. K. Gaye. In *The Complete Works of Aristotle,* edited by Jonathan Barnes. 2 vols. Princeton: Princeton University Press, 1984.

———. *Politics.* Translated by B. Jowett, in *The Complete Works of Aristotle* edited by Jonathan Barnes. 2 vols. Princeton: Princeton University Press, 1984.

Attfield, Robin. *A Theory of Value and Obligation.* London: Croom Helm, 1987.

Banfield, Edward C. *The Unheavenly City.* Boston: Little, Brown, 1968.

Barber, Benjamin. *Strong Democracy: Participatory Politics for a New Age.* Berkeley: University of California Press, 1984.

Barker, Ernest. *The Political Thought of Plato and Aristotle.* London: Methuen, 1906.

Barry, Brian. *The Liberal Theory of Justice.* Oxford: Clarendon Press, 1973.

———. *Political Argument.* London: Routledge & Kegan Paul, 1965.

Becker, Lawrence C. "Human Being: The Boundaries of the Concept." *Philosophy and Public Affairs* 4 (1974–75): 334–59.

Bell, Daniel. *The Cultural Contradictions of Capitalism.* New York: Basic Books, 1976.

Bentham, Jeremy. *An Introduction to the Principles of Morals and Legislation.* Edited by J. H. Burns and H. L. A. Hart. London: Methuen, 1970.

Block, N. J., and Gerald Dworkin. "IQ: Heritability and Inequality, Part 1." *Philosophy and Public Affairs* 3 (1973–74): 331–409.

Bosanquet, Bernard. "The Antithesis between Individualism and Socialism Philosophically

Considered.'' In *The Civilization of Christendom.* London: Swann Sonnenschein, 1893.
———. ''Character in Its Bearing on Social Causation.'' In *Aspects of the Social Problem,* edited by Bernard Bosanquet. London: Macmillan, 1895.
———. ''Luxury and Refinement.'' In *The Civilization of Christendom.* London: Swann Sonnenschein, 1893.
———. *The Philosophical Theory of the State.* London: Macmillan, 1899.
———. ''The Principle of Private Property.'' In *Aspects of the Social Problem,* edited by Bernard Bosanquet. London: Macmillan, 1895.
Bosanquet, Bernard, ed. *Aspects of the Social Problem.* London: Macmillan, 1895.
Bradley, F. H. *Appearance and Reality.* London: George Allen & Unwin, 1893.
———. *Essays on Truth and Reality.* Oxford: Clarendon Press, 1914.
———. *Ethical Studies.* 2d ed. Oxford: Clarendon Press, 1927.
———. ''The Limits of Individual and National Self-Sacrifice.'' In *Collected Essays.* Vol. 1. Oxford: Clarendon Press, 1935.
———. ''Mr. Sidgwick's Hedonism.'' In *Collected Essays.* Vol. 1. Oxford: Clarendon Press, 1935.
———. *The Principles of Logic.* 2d ed. London: Oxford University Press, 1922.
Brody, Baruch. *Identity and Essence.* Princeton: Princeton University Press, 1980.
Browning, Robert. *Poems of Robert Browning.* Edited by Donald Smalley. Boston: Houghton Mifflin, 1956.
Callahan, Tom. ''Masters of Their Own Game.'' *Time,* March 18, 1985: 38–50.
Clark, Stephen R. L. ''The Use of Man's Function in Aristotle.'' *Ethics* 82 (1972): 269–83.
Clarke, Peter. *Liberals and Social Democrats.* Cambridge: Cambridge University Press, 1978.
Cohen, G. A. *Karl Marx's Theory of History: A Defence.* Oxford: Clarendon Press, 1978.
———. ''Robert Nozick and Wilt Chamberlain: How Patterns Preserve Liberty.'' In *Justice and Economic Distribution,* edited by John Arthur and William H. Shaw. Englewood Cliffs, N.J.: Prentice-Hall, 1978.
Collini, Stefan. *Liberalism and Sociology: L. T. Hobhouse and Political Argument in England 1880–1914.* Cambridge: Cambridge University Press, 1979.
Darwin, Leonard. ''Memories of Down House.'' *The Nineteenth Century* 106 (1929): 118–23.
Davidson, Donald. *Essays on Actions and Events.* Oxford: Clarendon Press, 1980.
———. ''Mental Events.'' In *Essays on Actions and Events.* Oxford: Clarendon Press, 1980.
———. ''Psychology as Philosophy.'' In *Essays on Actions and Events.* Oxford: Clarendon Press, 1980.
Dworkin, Ronald. ''Liberalism.'' In *Public and Private Morality,* edited by Stuart Hampshire. Cambridge: Cambridge University Press, 1980.
Dyson, Freeman. *Infinite in All Directions.* New York: Harper & Row, 1988.
Elster, Jon. *Ulysses and the Sirens.* Cambridge: Cambridge University Press, 1979.
Finnis, John. *Natural Law and Natural Rights.* Oxford: Clarendon Press, 1980.
Foster, M. B. ''A Mistake of Plato's in the *Republic.*'' *Mind* 46 (1937): 386–93.
Frankena, William K. *Ethics.* 2d ed. Englewood Cliffs, N. J.: Prentice-Hall, 1973.
Freeden, Michael. *The New Liberalism: An Ideology of Social Reform.* Oxford: Clarendon Press, 1978.
Gardner, Howard. *Frames of Mind: The Theory of Multiple Intelligences.* New York: Basic Books, 1983.
Gibson, Mary. ''Rationality.'' *Philosophy and Public Affairs* 6 (1976–77): 193–225.
Glover, Jonathan. *What Sort of People Should There Be?* Harmondsworth: Penguin, 1984.

Gordon, Peter, and John White. *Philosophers as Educational Reformers: The Influence of Idealism on British Educational Thought and Practice*. London: Routledge & Kegan Paul, 1979.

Gould, Stephen Jay. *The Mismeasure of Man*. New York: Norton, 1981.

Green, Michael B., and Daniel Wilker. "Brain Death and Personal Identity." *Philosophy and Public Affairs* 9 (1979–80): 105–33.

Green, Thomas Hill. *Lectures on the Principles of Political Obligation*. In *Works of Thomas Hill Green*. Vol. 2, edited by R. L. Nettleship. London: Longmans, 1985–88.

———. "Liberal Legislation and Freedom of Contract." In *Works of Thomas Hill Green*. Vol. 3, edited by R. L. Nettleship. London: Longmans, 1985–88.

———. *Prolegomena to Ethics*. 5th ed. Oxford: Clarendon Press, 1907.

Greenwood, L. H. G. *Nicomachean Ethics Book Six*. Cambridge: Cambridge University Press, 1909.

Griffin, James. *Well-Being: Its Meaning, Measurement and Moral Importance*. Oxford: Clarendon Press, 1986.

Haksar, Vinit. *Equality, Liberty, and Perfectionism*. Oxford: Clarendon Press, 1979.

Hamilton, William. *Lectures on Metaphysics*. Vol. 1 of *Lectures on Metaphysics and Logic*. 2 vols. Edited by H. L. Mansel and John Veitch. New York: Sheldon and Company, 1880.

Hampshire, Stuart. *Thought and Action*. London: Chatto & Windus, 1960.

Hardy, G. H. *Ramanujan: Twelve Lectures on Subjects Suggested by His Life and Work*. Cambridge: Cambridge University Press, 1940.

Harman, Gilbert. "Practical Reasoning." *Review of Metaphysics* 29 (1976): 431–63.

Harrington, Michael. *The New American Poverty*. London: Firethorn, 1985.

———. *The Other America*. Baltimore: Penguin, 1963.

Hegel, G. W. F. *The Phenomenology of Spirit*. Translated by A. V. Miller. Oxford: Oxford University Press, 1977.

———. *The Philosophy of Right*. Translated by T. M. Knox. Oxford: Oxford University Press, 1967.

Hempel, Carl G. *Aspects of Scientific Explanation*. New York: Free Press, 1965.

Herrigel, Eugen. *Zen in the Art of Archery*. New York: Vintage, 1971.

Hobhouse, L. T. *The Elements of Social Justice*. London: George Allen & Unwin, 1922.

———. *Liberalism*. Oxford: Oxford University Press, 1911.

———. *The Rational Good*. London: George Allen & Unwin, 1921.

Hobson, John A. *Problems of Poverty*. London: Methuen, 1891.

Housman, A. E. *Collected Poems*. New York: Henry Holt, 1959.

Humboldt, Wilhelm von. *The Limits of State Action*. Translated by J. W. Burrow. Cambridge: Cambridge University Press, 1969.

Hurka, Thomas. "Average Utilitarianisms." *Analysis* 42 (1982): 65–9.

———. "Consequentialism and Content." *American Philosophical Quarterly* 29 (1992): 71–78.

———. "Perfectionism and Equality." In *On the Track of Reason: Essays in Honor of Kai Nielsen,* edited by Rodger Bechler, David Copp, and Béla Szabados. Boulder: Westview Press, 1992.

———. "Virtue as Loving the Good." *Social Philosophy and Policy* 9 (1992): 149–68.

———. "The Well-Rounded Life." *The Journal of Philosophy* 84 (1987): 707–26.

———. "Why Value Autonomy?" *Social Theory and Practice* 13 (1987): 361–82.

Jouvenel, Bertrand de. *The Ethics of Redistribution*. Cambridge: Cambridge University Press, 1951.

Kant, Immanuel. *The Critique of Practical Reason.* Translated by Lewis White Beck. Indianapolis: Bobbs-Merrill, 1956.

———. *The Doctrine of Virtue: Part 2 of The Metaphysics of Morals.* Translated by Mary J. Gregor. Philadelphia: University of Pennsylvania Press, 1964.

———. *Foundations of the Metaphysics of Morals.* Translated by Lewis White Beck. Indianapolis: Bobbs-Merrill, 1959.

———. *Religion Within the Limits of Reason Alone.* Translated by Theodore M. Greene and Hoyt H. Hudson. New York: Harper Torchbooks, 1960.

Kaufmann, Walter. *Nietzsche: Philosopher, Psychologist, Antichrist.* 4th ed. Princeton: Princeton University Press, 1974.

Kripke, Saul A. *Naming and Necessity.* Cambridge, Mass.: Harvard University Press, 1980.

Larmore, Charles E. *Patterns of Moral Complexity.* Cambridge: Cambridge University Press, 1987.

Leibniz, G. W. *The Leibniz-Arnauld Correspondence.* Translated by H. T. Mason. Manchester: Manchester University Press, 1967.

———. *Principles of Nature and of Grace.* Translated by Robert Latta. In *Leibniz: Selections,* edited by Phillip P. Wiener. New York: Scribner's, 1951.

———. *Theodicy.* Translated by E. M. Huggard. London: Routledge & Kegan Paul, 1951.

Lewis, Oscar. *The Children of Sanchez.* Harmondsworth: Penguin, 1964.

Lockwood, Michael. "Singer on Killing and the Preference for Life." *Inquiry* 22 (1979): 157–70.

MacIntyre, Alasdair. *After Virtue.* London: Duckworth, 1981.

Macpherson, C. B. *Democratic Theory: Essays in Retrieval.* Oxford: Clarendon Press, 1973.

McTaggart, John McTaggart Ellis. *The Nature of Existence.* 2 vols. Cambridge: Cambridge University Press, 1921.

———. *Studies in Hegelian Cosmology.* Cambridge: Cambridge University Press, 1901.

Malinovitch, Stanley. "Knowledge and Evaluation." *Canadian Journal of Philosophy* 4 (1974): 79–95.

Marx, Karl. *Capital.* Vol. 1. Translated by Ben Fowkes. New York: Vintage, 1977.

———. *Critique of the Gotha Programme* (excerpt). In *Karl Marx: Selected Writings,* edited by David McLellan. Oxford: Oxford University Press, 1977.

———. *Economic and Philosophical Manuscripts* (excerpt). In *Karl Marx: Selected Writings,* edited by David McLellan. Oxford: Oxford University Press, 1977.

———. *Grundrisse.* Translated by M. Nicolaus. Harmondsworth: Penguin, 1973.

Marx, Karl and Friedrich Engels. *The Communist Manifesto.* In *Karl Marx: Selected Writings,* edited by David McLellan. Oxford: Oxford University Press, 1977.

———. *The German Ideology* (excerpt). In *Karl Marx: Selected Writings,* edited by David McLellan. Oxford: Oxford University Press, 1977.

———. *The Holy Family* (excerpt). In *Karl Marx: Selected Writings,* edited by David McLellan. Oxford: Oxford University Press, 1977.

Mill, John Stuart. *Considerations on Representative Government.* In *Utilitarianism, Liberty, and Representative Government.* London: J. M. Dent, 1910.

———. *On Liberty.* In *Utilitarianism, Liberty, and Representative Government.* London: J. M. Dent, 1910.

Moore, G. E. *Principia Ethica.* Cambridge: Cambridge University Press, 1903.

Murray, Charles. *Losing Ground: American Social Policy 1950–1980.* New York: Basic Books, 1984.

Nagel, Thomas. *Equality and Partiality.* New York: Oxford University Press, 1991.

———. "The Fragmentation of Value." In *Mortal Questions.* Cambridge: Cambridge University Press, 1979.

———. "Moral Conflict and Political Legitimacy." *Philosophy and Public Affairs* 16 (1987): 215–40.

———. *The View From Nowhere*. New York: Oxford University Press, 1986.

Nehamas, Alexander. *Nietzsche: Life as Literature*. Cambridge, Mass.: Harvard University Press, 1985.

Nettleship, R. L. *Lectures on the Republic of Plato*. London: Macmillan, 1925.

Nielsen, Kai. "Alienation and Self-Realization." *Philosophy* 48 (1973): 21–33.

———. *Equality and Liberty: A Defense of Radical Egalitarianism*. Totowa, N.J.: Rowman & Allenheld, 1985.

Nietzsche, Friedrich. *The Antichrist*. Translated by Walter Kaufmann. In *The Portable Nietzsche,* edited by Walter Kaufmann. Harmondsworth: Penguin, 1976.

———. *Beyond Good and Evil*. Translated by Walter Kaufmann. New York: Vintage, 1966.

———. *The Gay Science*. Translated by Walter Kaufmann. New York: Vintage, 1974.

———. *Mixed Opinions and Maxims* (excerpt). Translated by Walter Kaufmann. In *On the Genealogy of Morals/Ecce Homo,* translated by Walter Kaufmann and R. J. Hollingdale. New York: Vintage, 1969.

———. *On the Genealogy of Morals*. Translated by Walter Kaufmann and R. J. Hollingdale; *Ecce Homo*. Translated by Walter Kaufmann. New York: Vintage, 1969.

———. *Schopenhauer as Educator*. Translated by James W. Hillesheim and Malcolm R. Simpson. South Bend, Ind.: Gateway Editions, 1965.

———. *Thus Spoke Zarathustra*. Translated by Walter Kaufmann. In *The Portable Nietzsche,* edited by Walter Kaufmann. Harmondsworth: Penguin, 1976.

———. *Twilight of the Idols*. Translated by Walter Kaufmann. In *The Portable Nietzsche,* edited by Walter Kaufmann. Harmondsworth: Penguin, 1976.

———. *The Use and Abuse of History*. Translated by Adrian Collins. Indianapolis: Bobbs-Merrill, 1949.

———. *The Will to Power*. Translated by Walter Kaufmann and R. J. Hollingdale. New York: Vintage, 1968.

Norton, David L. *Personal Destinies: A Philosophy of Ethical Individualism*. Princeton: Princeton University Press, 1976.

Nozick, Robert. *Anarchy, State, and Utopia*. New York: Basic Books, 1974.

———. *Philosophical Explanations*. Cambridge, Mass.: Harvard University Press, 1981.

Parfit, Derek. "Overpopulation and the Quality of Life." In *Applied Ethics,* edited by Peter Singer. Oxford: Oxford University Press, 1986.

———. *Reasons and Persons*. Oxford: Clarendon Press, 1984.

Passmore, John. *The Perfectibility of Man*. New York: Scribner's, 1970.

Plato. *Euthydemus*. Translated by W. H. D. Rouse. In *The Collected Dialogues of Plato,* edited by Edith Hamilton and Huntington Cairns. Princeton: Bollingen, 1961.

———. *Laws*. Translated by A. E. Taylor. In *The Collected Dialogues of Plato,* edited by Edith Hamilton and Huntington Cairns. Princeton: Bollingen, 1961.

———. *Philebus*. Translated by R. Hackforth. In *The Collected Dialogues of Plato,* edited by Edith Hamilton and Huntington Cairns. Princeton: Bollingen, 1961.

———. *Republic*. Translated by G. M. A. Grube. Indianapolis: Hackett, 1974.

Prichard, H. A. "The Meaning of *Agathon* in the *Ethics* of Aristotle." *Philosophy* 10 (1935): 27–39.

Putnam, Hilary. "Is Semantics Possible?" In *Mind, Language, and Reality: Philosophical Papers*. Vol. 2. Cambridge: Cambridge University Press, 1975.

———. "The Meaning of 'Meaning'." In *Mind, Language, and Reality: Philosophical Papers*. Vol. 2. Cambridge: Cambridge University Press, 1975.

Rashdall, Hastings. *The Theory of Good and Evil.* 2 vols. London: Oxford University Press, 1907.

Rawls, John. "The Independence of Moral Theory." *Proceedings and Addresses of the American Philosophical Association* 48 (1974–75): 5–22.

———. "Justice as Fairness: Political not Metaphysical." *Philosophy and Public Affairs* 14 (1985): 223–51.

———. "Kantian Constructivism in Moral Theory." *The Journal of Philosophy* 77 (1980): 515–72.

———. *A Theory of Justice.* Cambridge, Mass.: Harvard University Press, 1971.

Regan, Donald H. *Utilitarianism and Co-operation.* Oxford: Clarendon Press, 1980.

Richards, David A. J. "Justice and Equality." In *And Justice for All,* edited by Tom Regan and Donald VanDeVeer. Totowa, N.J.: Rowman and Littlefield, 1982.

Ross, W. D. *The Right and the Good.* Oxford: Clarendon Press, 1930.

Schor, Juliet B. *The Overworked American: The Unexpected Decline of Leisure.* New York: Basic Books, 1991.

Scruton, Roger. *The Meaning of Conservatism.* Harmondsworth: Penguin, 1980.

Sen, Amartya K. *Collective Choice and Social Welfare.* San Francisco: Holden-Day, 1970.

Sidgwick, Henry. *Lectures on the Ethics of T. H. Green, Mr. Herbert Spencer, and J. Martineau.* London: Macmillan, 1902.

———. *The Methods of Ethics.* 7th ed. London: Macmillan, 1907.

Sikora, R. I. "Is It Wrong to Prevent the Existence of Future Generations?" In *Obligations to Future Generations,* edited by R. I. Sikora and Brian Barry. Philadelphia: Temple University Press, 1978.

Slote, Michael. *Common-Sense Morality and Consequentialism.* London: Routledge & Kegan Paul, 1985.

Spinoza, Benedict de. *Ethics.* In *The Chief Works of Benedict de Spinoza.* Vol. 2. Translated by R. H. M. Elwes. New York: Dover, 1955.

Sturt, George. *The Wheelwright's Shop.* Cambridge: Cambridge University Press, 1976.

Suits, Bernard. *The Grasshopper: Games, Life, and Utopia.* Toronto: University of Toronto Press, 1978.

Swift, Jonathan. *Gulliver's Travels.* London: Heinemann, 1964.

Tawney, R. H. *Equality.* London: George Allen & Unwin, 1964.

Taylor, Charles. "Marxism and Empiricism." In *British Analytical Philosophy,* edited by Bernard Williams and Alan Montefiore. London: Routledge & Kegan Paul, 1966.

Unger, Roberto Mangabeira. *Knowledge and Politics.* New York: Free Press, 1975.

van Fraassen, B. C. "Presuppositions, Supervaluations, and Free Logic." In *The Logical Way of Doing Things,* edited by Karel Lambert. New Haven: Yale University Press, 1969.

Vendler, Zeno. "The Grammar of Goodness." *The Philosophical Review* 72 (1963): 446–65.

Wiggins, David. *Sameness and Substance.* Oxford: Basil Blackwell, 1980.

Williams, Bernard. *Ethics and the Limits of Philosophy.* Cambridge, Mass: Harvard University Press, 1985.

———. *Morality: An Introduction to Ethics.* New York: Harper & Row, 1972.

———. "Persons, Character, and Morality." In *Moral Luck.* Cambridge: Cambridge University Press, 1981.

Xenophon. *Apology.* Translated by O. J. Todd. In *Memorabilia, Oeconomicus, Symposium, Apology.* London: Heinemann, 1968.

———. *Memorabilia.* Translated by E. C. Marchant. In *Memorabilia, Oeconomicus, Symposium, Apology.* London: Heinemann, 1968.

Index